Use R!

Series Editors

Robert Gentleman
Kurt Hornik
Giovanni Parmigiani

More information about this series at http://www.springer.com/series/6991

David Magis • Duanli Yan • Alina A. von Davier

Computerized Adaptive and Multistage Testing with R

Using Packages *catR* and *mstR*

 Springer

David Magis
Department of Education
University of Liege
Liege, Belgium

Duanli Yan
Educational Testing Service
Princeton, NJ, USA

Alina A. von Davier
ACTNext by ACT
Iowa City, IA, USA

ISSN 2197-5736   ISSN 2197-5744   (electronic)
Use R!
ISBN 978-3-319-88735-7   ISBN 978-3-319-69218-0   (eBook)
https://doi.org/10.1007/978-3-319-69218-0

Printed on acid-free paper

This Springer imprint is published by Springer Nature
The registered company is Springer International Publishing AG
The registered company address is: Gewerbestrasse 11, 6330 Cham, Switzerland

To Maxime, Éléonore, and Anaëlle,
probably future R users

David Magis

To my daughter Victoria Song, an avid
R programmer

Duanli Yan

To my son, Thomas, whose time has come
to learn R

Alina A. von Davier

Foreword

Adaptive testing is, by now, a well-established procedure in educational and psychological assessment. Early adopters in the educational field in the USA were the licensing exam (NCLEX/CAT) of the National Council of State Boards of Nursing and the Graduate Record Examination (GRE). But at the time of this writing, the trend of using adaptive testing is worldwide. For instance, in the Netherlands, several adaptive tests are available for the compulsory nationwide final assessments in primary education. Further, also large-scale international educational surveys have implemented adaptive assessments; the Organisation for Economic Co-operation and Development (OECD) Program for International Student Assessment (PISA) and Program for the International Assessment for Adult Competencies (PIAAC) are prominent examples. Adaptive testing has also reached fields beyond educational assessment, such as epidemiology (the Patient-Reported Outcomes Measurement Information System (PROMIS) project for health assessment is an example), and organizational assessment (for supporting selection and promotion decisions). Most of these applications are based on adaptive selection of individual items. A more recent development is towards adaptive selection of testlets. This development is mainly motivated by the ease with which content control can be supported and generally goes under the name of multistage testing.

Given its widespread use, it may come as no surprise that good introductions to adaptive testing are available. I mention two edited volumes: *Computerized Adaptive Testing: A Primer* edited by Howard Wainer, which gives an excellent relatively nontechnical introduction to the key ingredients and practical implementation issues of adaptive testing, and *Elements of Adaptive Testing* edited by van der Linden and Glas, which delves a bit deeper into the statistical issues involved. An excellent introduction to multistage testing can be found in *Computerized Multistage Testing: Theory and Applications* edited by Yan, von Davier, and Lewis.

So where is the present volume and the software that goes with it to be positioned? First of all, the book provides a general overview of the statistical technology underlying adaptive testing, together with R packages that can provide illustrations for the theory. But there is more. When developing an adaptive test, many questions must be answered and many decisions must be made. For instance, is my item bank big enough given the foreseen number of respondents and lifespan of the test, is the test length appropriate for the targeted reliability, and are the item parameters adequate for the targeted population, to mention a few questions. The website of the International Association for Computerized Adaptive Testing (IACAT) adds some more issues: content balancing (to create a test where the content matter is appropriately balanced), item exposure control (to prevent compromising of the test), combining multiple scales, and proctoring of administration via the Internet. To complicate things, all these aspects usually interact. The simulation studies such as those supported by the R packages can help answering such questions and support the decisions. The choice of R is a good one, motivated by the fact that it has become the most versatile and flexible open-source platform for statistical analyses and creation of graphical output. For most developers and users of statistical tools, R has become the standard of industry. It offers practitioners the opportunity to add their own functionality to existing applications and provides a common ground for exchanging their software. Therefore, I am sure that this book and the simulation software accompanying it will be extremely helpful for designing adaptive tests.

Professor of Social Science Research Methodology   Cees Glas
University of Twente, The Netherlands

Acknowledgments

This book takes its origin from an inspiring discussion between the authors during the meeting of the International Association for Computerized Adaptive Testing (IACAT) in Cambridge (UK), October 2015. From this early project, several drafts were written, re-organized, reviewed, and updated to end up with the current version of the book.

The authors wish to express their sincere acknowledgments to the many people who took some of their precious time to read the provisional chapters and make insightful comments: Hong Jiao (University of Maryland), Kim Fryer, Lixiong Gu, Sooyeon Kim, Yanming Jiang, Longjuan Liang, Guangming Ling, Yuming Liu, Manfred Steffen, Fred Robin, Jonathan Weeks, Meng Wu (Educational Testing Service). The authors want to express their gratitude to Cees Glas who accepted to preface this book long before it came out in its final version. Eventually, special thanks to Jim Carlson from Educational Testing Service for his comments and suggestions throughout the whole book, and to Andrew Cantine (technical editor, ACTNext) for carefully editing the writing of this book.

Contents

Acronyms

CAT   Computerized Adaptive Testing
IRT   Item Response Theory
MST   Multistage Testing

List of Figures

List of Tables

Chapter 1
Overview of Adaptive Testing

In this book, we provide a general overview of computerized adaptive tests (CAT) and multistage tests (MST), as well as item response theory (IRT). We briefly discuss the important concepts associated with CATs and MSTs. We then introduce the R open source software packages **catR** and **mstR** for CAT and MST simulations, and provide examples on how to use them.

The book is structured to take the reader through all key aspects of CAT and MST simulations, from data generation, item banking, item parameter generation for simulations, to estimation and scoring by using **catR** and **mstR** functions. This introductory chapter also provides a comparison of the common forms of tests and an overview of subsequent chapters.

1.1 Linear Test, CAT and MST

Traditionally, linear tests have been the most common way of measuring test takers' knowledge, skills, and abilities, especially in educational assessments. However, over the last two decades, computer science and technology have advanced rapidly and the demand for computer-based tests (CBT) has greatly increased. In particular, computerized adaptive tests have been used in many real-world settings, due to their efficiency and precision in testing. More recently, computerized multistage tests (MST) have become very popular for their features and efficiency.

1.1.1 Linear Test

A *linear test* is usually administered on paper and is commonly referred to as a paper-and-pencil test. In a linear test, all test takers are presented with every item

© Springer International Publishing AG 2017
D. Magis et al., *Computerized Adaptive and Multistage Testing with R*, Use R!,
https://doi.org/10.1007/978-3-319-69218-0_1

regardless of whether the items are very easy or very difficult for them. Test takers with a relatively high (or low) level of knowledge or ability would still need to answer the easy (or difficult) items. Because it is likely that low ability test takers will get all difficult items wrong and high ability test takers will get all easy items correct, these easy and difficult items contribute little information toward measuring test takers' abilities at the higher and lower ends of measurement scale, respectively. The result is that linear tests require large numbers of items in order to obtain uniformly good precision in the final scores (Rudner, 1998).

A *computer-based test* is a test administered on a computer and can be a linear, adaptive, or multistage test. A linear CBT is similar to a traditional, linear, paper-and-pencil test, except that it is administered via computer. Therefore, a CBT enjoys many of the advantages associated with computerization, such as the flexibility of test scheduling and the efficiency of test administration, test assembly, and score reporting, but suffers the same limitations as a linear paper-and-pencil test.

1.1.2   CAT

A CAT is also a test administered on a computer. In addition to computerized administration, it uses an algorithm-based approach to administer test items. Specifically, the items that are chosen and administered are adapted to the test taker's estimated ability level during the testing process, the estimated ability is continuously updated after each item is administered. Thus, a CAT is an item-level adaptive test and can be of fixed or variable length. The ability estimate is used not only to represent a test taker's ability level, but also to determine the selection of subsequent items from an available item bank. Typically, a CAT can be much more efficient (e.g., shorter test length) than a traditional linear test (Wainer, Kaplan, & Lewis, 1992). Therefore, CATs have been widely used in recent years.

CATs have been shown to have advantages over traditional linear testing. These include more efficient and precise measurement of test takers' performances across the entire ability distribution (Hendrickson, 2007; Lord, 1980; Wainer, 2000). As mentioned earlier, traditional linear tests measure test takers of average ability within the group quite well, the highest precision often occurring for the scores of test takers whose ability is average for the intended measurement group. For test takers near the extremes of the measurement scale, linear tests give less precise measurements, when the test is relatively short. In other words, linear tests have difficulty providing precise measurements for the whole range of ability levels present within the group (Betz & Weiss, 1974; Hambleton & Swaminathan, 1985; Hendrickson, 2007; Lord, 1980). Since CATs focus measurement on an individual test taker's estimated ability level, they can provide precise measurement for all test takers, including those of average ability and those near the ends of the measurement scale (Hendrickson, 2007; Lord, 1974; Mills, Potenza, Fremer, & Ward, 2002; Wainer et al., 1992). Figure 3.1 in Chap. 3 is a schematic of a CAT.

1.1.3 MST

An MST also involves adaptation to test taker's estimated ability level. It is a compromise between a CAT and a linear test, with features from both designs. Based on decades of research (Betz & Weiss, 1974; Hambleton & Swaminathan, 1985; Hendrickson, 2007; Lord, 1980; Luecht, 1998; van der Linden, Ariel, & Veldkamp, 2006; van der Linden & Glas, 2010; Wainer, Bradlow, & Wang, 2007; Yan, von Davier, & Lewis, 2014), MSTs can incorporate most of the advantages from CATs and linear tests, while minimizing their disadvantages. Thus, MSTs are becoming more and more popular in real world testing applications.

While MSTs are very similar to CATs, rather than administering test items one at a time adaptively for each test taker with a typical CAT, the algorithm administers pre-assembled groups of items (called *modules*) for each test taker, and the test is built up in *stages* (see Fig. 6.1 in Chap. 6). In an MST, all test takers are administered an initial set of items, usually called a routing test, at the first stage of testing (module A in Fig. 6.1). Based on their performance, test takers are routed to one of several different modules adapted to the test taker's estimated ability level (modules B and C in Fig. 6.1) at the second stage. The number of stages and number of modules available per stage can vary, depending on the design of the MST. The final stage of an MST's module is often called a measurement test (modules D, E and F in Fig. 6.1). Similar to CATs, MSTs also focus measurement on an individual test taker's ability level, so they can provide precise measurement for all test takers including those of average ability and those near the ends of the measurement scale (Yan et al., 2014).

Table 1.1 lists major advantages and disadvantages of linear tests, CATs and MSTs.

1.2 Organization of This Book

This book is organized as follows. First, Chap. 2 provides an overview of item response theory (IRT) from principles and assumptions to commonly used IRT models, parameter and ability estimation, and discussions thereof. The remainder of the book is split into two main parts, the first one devoted to CAT and the second to MST.

Chapter 3 provides an overview of CAT from the basics, to test designs and implementations, IRT-based CAT item banking, assembly, estimation, scoring, linking and equating, as well as the tree-based CAT. Chapter 4 introduces the R open source package **catR** and its functions. It gives detailed illustrations on how-to use **catR** functions for each step of a CAT simulation including data generation, parameter generation, and scoring. Chapter 5 demonstrates some examples of simulations using **catR** to help readers and **catR** users in practice.

Table 1.1 A comparison of linear, CAT, and MST designs

Type of test	Advantages	Disadvantages
Linear test	Ease of assembly	Full length test
	Ease of administration	Inefficient for measurement
	Least effort for test development (TD)	Inflexible test schedule for test takers
		Prone to test copying
CAT	Shorter test length	Complicated to implement
	Efficient for measurement	Depends on strong model
	Flexible test schedule for test takers	assumptions
	Avoids test copying	Requires a large calibration data set
		Greatest effort for TD
		Item exposure more difficult to control
		Costly to administer via computer
		Robustness concerns
MST	Intermediate test length	Depends on model assumptions
	Efficient for measurement	Longer than CAT but shorter than
	Allows test taker item review (within	linear test
	modules)	Item exposure concerns (similar to
	Easier to implement	CAT)
	Easier to assemble	Costly to administer via computer
	Moderate effort for TD	(similar to CAT)
	Flexible test schedule for test takers	
	Reduces test copying	

In the second part of this book, Chap. 6 provides an overview of MST from the basics, to test designs and implementations, IRT-based MST item banking, assembly, estimation, scoring, linking and equating, as well as a tree-based MST. Chapter 7 presents the R package **mstR**, its specificities and similarities with respect to **catR**. It gives detailed illustrations on how to use **mstR** functions for each step of an MST simulation. Finally, Chap. 8 provides examples of simulations using **mstR**.

The R packages **catR** and **mstR** offer a variety of options for generating response patterns in a simulated CAT or MST environment by first providing a pre-calibrated item bank, then by providing options according to the CAT or MST assessment design. The packages integrate several rules for early item or module selection, ability estimation, next item selection or module selection, stopping criteria, and can control for crucial issues such as item exposure and content balancing for CAT and MST. The packages' general architecture makes **catR** and **mstR** flexible, easy to update, and many of the components can even be used outside the CAT and MST framework (for instance, the ability estimation and related standard error computation routines). The packages are also flexible with respect to the choice of the underlying logistic item response model and can handle a large number of pattern generations for given true ability levels. This provides the advantage of being able to evaluate the quality of item banks by generating a large number of response patterns under various settings (different ability levels, different item selection rules, etc.) and comparing the results to the simulated patterns. The packages can also be used as a computational engine for real CAT and MST assessment platforms, such as the web-based platform *Concerto* (Kosinski et al., 2013).

In this book, version 3.12 of **catR** and version 1.0 of **mstR** are considered. As open-source packages, they can be easily updated or modified by R users. The packages are available from the Comprehensive R Archive Network at http://CRAN. R-project.org.

Chapter 2
An Overview of Item Response Theory

An important component of both CAT and MST is the use of item response theory (IRT) as an underlying framework for item bank calibration, ability estimation, and item/module selection. In this chapter, we present a brief overview of this theory, by providing key information and introducing appropriate notation for use in subsequent chapters. Only topics and content directly related to adaptive and multistage testing will be covered in this chapter; appropriate references for further reading are therefore also mentioned.

2.1 Principles and Assumptions of Item Response Theory

Item response theory is a field of psychometrics that focuses on the measurement and evaluation of psychological or educational *latent traits*, i.e., unobservable characteristics such as ability, intelligence or competence levels. It takes its origin in the pioneering works of Lord (1951), Green (1950), Rasch (1960) and Lord and Novick (1968), and has received increasing interest since the 1980s, with reference books written by Lord (1980), Hambleton and Swaminathan (1985), Linden and Hambleton (1997), Embretson and Reise (2000), Baker and Kim (2004), De Boeck and Wilson (2004), Yen and Fitzpatrick (2006), and Rao and Sinharay (2007), van der Linden and Hambleton (2017) among others. Of primary interest for this book, IRT has also been considered as an appropriate framework for (computerized) adaptive testing and is discussed in detail, e.g., Weiss (1983), Wainer (2000) and Linden and Glas (2010).

In practical terms, IRT can be seen as a framework for evaluating or estimating latent traits by means of *manifest* (observable) variables and appropriate (statistical) *psychometric* models. More precisely, the latent traits (hereafter referred to as *ability levels*) are evaluated by administering a set of *items* or questions to the test takers. These items are designed to target and measure the abilities as precisely and accu-

© Springer International Publishing AG 2017
D. Magis et al., *Computerized Adaptive and Multistage Testing with R*, Use R!,
https://doi.org/10.1007/978-3-319-69218-0_2

rately as possible. Test takers' responses to these items are the manifest variables and can be of various types: dichotomous (e.g., true/false), polytomous (e.g., Likert scales such as "totally disagree", "disagree", "agree" and "totally agree"), multiple-choice, and so on. The information collected by the test administrations can be used to estimate the underlying latent traits of interest by fitting an appropriate IRT model and by making use of specific *item characteristics*.

In other words, an IRT model can be seen as a mathematical function, called the *item response function* (IRF) or *item characteristic curve* (ICC) that describes the probability of observing each possible item response, as a function of a set of person's ability levels and item's parameters:

$$P_{jk}(\boldsymbol{\theta}_i, \boldsymbol{p}_j) = Pr(X_{ij} = k | \boldsymbol{\theta}_i, \boldsymbol{p}_j) = f(k, \boldsymbol{\theta}_i, \boldsymbol{p}_j) \tag{2.1}$$

In Eq. (2.1), $\boldsymbol{\theta}_i$ is a multivariate vector of latent traits for test taker i ($i = 1, \dots, I$); X_{ij} is the response of test taker i to item j ($j = 1, \dots, J$); k is one of the possible item responses; and $\boldsymbol{p}_j$ is the set of item parameters. For dichotomously scored items, $k \in \{0, 1\}$ where response 0 stands for an incorrect response and 1 for a correct response. In case of polytomously scored items, it will be assumed that X_{ij} can take values in $\{0, \dots, K_j\}$, that is, among $K_j + 1$ possible response categories (not necessarily being ordered).

Historically, dichotomous IRT models were built upon three main assumptions (Hambleton & Swaminathan, 1985):

1. *Unidimensionality of the latent trait*: each item targets one and only one latent trait, or in other words the multidimensional vector $\boldsymbol{\theta}_i$ is reduced to a single latent trait θ_i.
2. *Local (item) independence*: at a given vector of ability levels $\boldsymbol{\theta}_i$, the responses $\{X_{ij} : j = 1, \dots, J\}$ to the J test items are stochastically independent.
3. *Monotonicity of the IRF*: the response probability (2.1) is monotonically increasing or decreasing with the ability level.

This set of assumptions allowed the construction and application of well-known dichotomous IRT models (Sect. 2.2.1) such as the Rasch model (Rasch, 1960) and logistic IRT models (Lord, 1980).

However, with the increase in complexity of test designs, item structures, data collection processes and computational resources, more complex models emerged that do not rely on all of the assumptions stated above. For instance, *multidimensional IRT models* (Sect. 2.2.3) involve vectors of correlated ability levels $\boldsymbol{\theta}_i$ that can be estimated. It is plausible to assume that some items may actually target more than one latent trait, or some tests may be built with a subset of items aimed at measuring different trait levels (e.g., large-scale assessment surveys like PISA, in which domains such as mathematics, reading and science are jointly estimated from booklets of items). Moreover, some test designs specifically involve items whose responses cannot be assumed to be locally independent, e.g., *testlets* (small sets of items that relate to a common stimulus or depend on each other by construction). It is therefore most likely that responses to testlet items will be correlated to some extent. Specific IRT models, referred to as *testlet IRT models* (Sect. 2.2.4.1), were developed expressly for that purpose (Wainer, Bradlow, & Wang, 2007). Finally, polytomous IRT models (Sect. 2.2.2) do not make use of the monotonicity of the IRFs for every possible item responses, as will be illustrated further.

2.2 Commonly Used IRT Models

In this section, we briefly outline the most commonly used IRT models. Unidimensional models (for both dichotomous and polytomous data) are presented first. Their multidimensional extensions, as well as some specific IRT models are also briefly mentioned for completeness.

2.2.1 *Unidimensional Dichotomous IRT Models*

Let us start with the simplest framework of unidimensional models for dichotomously scored items, that is $X_{ij} \in \{0, 1\}$ and $\boldsymbol{\theta}_i = \theta_i$. The IRF (2.1) reduces to

$$P_{j1}(\theta_i, \boldsymbol{p}_j) = Pr(X_{ij} = 1 | \theta_i, \boldsymbol{p}_j) = P_j(\theta_i, \boldsymbol{p}_j) \tag{2.2}$$

and represents the probability of a correct response, while

$$P_{j0}(\theta_i, \boldsymbol{p}_j) = Pr(X_{ij} = 0 | \theta_i, \boldsymbol{p}_j) = 1 - P_j(\theta_i, \boldsymbol{p}_j) = Q_j(\theta_i, \boldsymbol{p}_j) \tag{2.3}$$

represents the probability of an incorrect response.

Two mathematical functions used to characterize the IRFs (2.2) and (2.3) were suggested: the *standard normal probability* distribution function, leading to the so-called *normal ogive model* (Hambleton & Swaminathan, 1968; Lord & Novick, 1985), and the *logistic* distribution function, leading to so-called *logistic IRT models*. The latter became more popular than the former, primarily because of ease of computation (at a time when computational resources were more limited than at present), and are still commonly used nowadays in practical assessments. We will restrict our discussion to logistic IRT models in this presentation.

The simplest logistic IRT model is called the *one-parameter logistic model* (1PL; Lord & Novick, 1968), also sometimes referred to as the *Rasch model* (Rasch, 1960). It involves a single item parameter b_j that represents its difficulty level. The corresponding IRF (2.2) takes the following form:

$$P_j(\theta_i, \boldsymbol{p}_j) = Pr(X_{ij} = 1 | \theta_i, b_j) = \frac{\exp(D\,[\theta_i - b_j])}{1 + \exp(D\,[\theta_i - b_j])}. \tag{2.4}$$

The constant D is a scaling constant that was introduced by Lord and Novick (1968). In order to match the logistic and probit metrics very closely, D is fixed to 1.702 (Haley, 1952). Fixing D to 1, as is commonly done nowadays, yields the logistic metric and greatly simplifies the model equations. We will therefore apply this convention and not mention the D constant anymore in this chapter.

Equation (2.4) specifies that the probability of a correct response decreases when the difficulty level increases (at fixed ability levels) and takes the value of 0.5 when the ability level exactly matches the difficulty of the item.

Extensions of the 1PL model were introduced to add more item characteristics (Lord, 1980; Lord & Novick, 1968). First, the *two parameter logistic* (2PL) model introduces the discrimination level a_j of the item, that is, it allows the IRFs to have different slopes across items:

$$P_j(\theta_i, \boldsymbol{p}_j) = Pr(X_{ij} = 1 | \theta_i, a_j, b_j) = \frac{\exp\left[a_j\left(\theta_i - b_j\right)\right]}{1 + \exp\left[a_j\left(\theta_i - b_j\right)\right]} \tag{2.5}$$

The *three-parameter logistic* (3PL) model includes the addition of the lower asymptote (sometimes called pseudo-guessing) parameter c_j, allowing the IRF to asymptote towards some positive probability (instead of zero as in 1PL and 2PL models; Birnbaum, 1968):

$$P_j(\theta_i, \boldsymbol{p}_j) = Pr(X_{ij} = 1 | \theta_i, a_j, b_j, c_j) = c_j + (1 - c_j)\frac{\exp\left[a_j\left(\theta_i - b_j\right)\right]}{1 + \exp\left[a_j\left(\theta_i - b_j\right)\right]} \tag{2.6}$$

Eventually, the *four-parameter logistic* (4PL) model assumes an upper asymptote parameter d_j that allows maximal probability to be lower than one, and is subsequently sometimes referred to as the inattention parameter (Barton & Lord, 1981):

$$P_j(\theta_i, \boldsymbol{p}_j) = Pr(X_{ij} = 1 | \theta_i, a_j, b_j, c_j, d_j) = c_j + (d_j - c_j)\frac{\exp\left[a_j\left(\theta_i - b_j\right)\right]}{1 + \exp\left[a_j\left(\theta_i - b_j\right)\right]} \tag{2.7}$$

In other words, each logistic IRT model considers a specific set of item parameters: $\boldsymbol{p}_j = \{b_j\}$ for the 1PL model; $\boldsymbol{p}_j = \{a_j, b_j\}$ for the 2PL model; $\boldsymbol{p}_j = \{a_j, b_j, c_j\}$ for the 3PL model; and $\boldsymbol{p}_j = \{a_j, b_j, c_j, d_j\}$ for the 4PL model.

Examples of these four logistic IRT models are illustrated in Fig. 2.1 by varying each item parameter that is specific to each model. The 1PL model (upper left panel) illustrates three increasing difficulty levels, and clearly highlights that an increase in difficulty yields a shift in the IRF to the right side of the ability scale and hence, lower probabilities of correct responses at a given ability level. For the 2PL model (upper right panel), several discrimination values are represented. At larger discrimination levels, the IRF gets steeper and a small increase in ability around the item difficulty level leads to a notable increase in probability (that is, the test item discriminates better between low and high ability test takers). Conversely, lower discrimination levels yield flatter IRFs and such items do not allow for easily

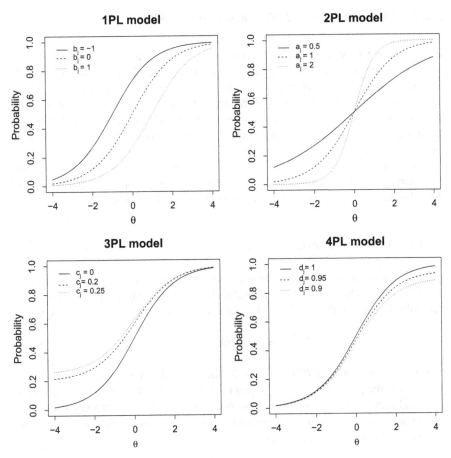

Fig. 2.1 Item response functions of various unidimensional logistic IRT models. For the 2PL, 3PL and 4PL models, the difficulty b_j is fixed to zero. For the 3PL and 4PL models, the discrimination a_j is fixed to one. For the 4PL model, the lower asymptote c_j is fixed to zero

discrimination between different levels of ability. The 3PL model (lower left panel) presents increasing lower-tail probabilities as the lower asymptote increases. Said differently, even low-ability test takers get a non-zero probability of answering the item correctly (for instance, this is the case with multiple-choice items). Finally, the 4PL model (lower right panel) can model upper-tail probabilities with a value lower than one. In other words, even high ability test takers can answer the item incorrectly, which may happen, for instance, when the test taker is subject to external phenomena, like tiredness or inattention.

2.2.2 Unidimensional Polytomous IRT Models

Items may not produce a simple binary response (i.e., true/false) but offer more choices, for instance responses graded on a Likert scale (from "Totally disagree" to "Totally agree") or multiple-choice items. For the latter, one always has the option of recoding the responses as binary outcomes, but at the cost of loosing specific distractor information. In the former case such an outcome reduction is often impossible since there is no "correct'" response. Thus, more complex models involving polytomous responses are required.

There are two main classes of (unidimensional) polytomous IRT models (Thissen & Steinberg, 1986): *difference models* and *divide-by-total models*. Both classes are described hereafter and their main IRT models are briefly outlined.

2.2.2.1 Difference Models

Difference models are defined by first setting an appropriate mathematical form to *cumulative probabilities*, that is, the probability that the item response will be one particular k value (with $k \in \{0, \ldots, K_j\}$) or any "larger" value (i.e. $k + 1$ to K_j):

$$P_{jk}^*(\theta_i, \pmb{p}_j) = Pr(X_{ij} \geq k | \theta_i, \pmb{p}_j), \tag{2.8}$$

with the assumption that $P_{jk}^*(\theta_i, \pmb{p}_j) = 0$ when $k > K_j$. It is straightforward that with $k = 0$, the probability (2.8) is equal to one, and when $k = K_j$ the probabilities (2.1) and (2.8) are identical. With these notations, the category probabilities (2.1) can be recovered from cumulative probabilities (2.8) by appropriate differences of successive cumulative probabilities:

$$P_{jk}(\theta_i, \pmb{p}_j) = P_{jk}^*(\theta_i, \pmb{p}_j) - P_{j,k+1}^*(\theta_i, \pmb{p}_j) \tag{2.9}$$

for any $k \in \{0, \ldots, K_j\}$.

The best known difference model is the *graded response model* (GRM; Samejima, 1969). Basically, it consists of defining the cumulative probabilities (2.8) as logistic functions from a dichotomous 2PL model:

$$P_{jk}^*(\theta_i, \pmb{p}_j) = \frac{\exp\left[\alpha_j\left(\theta_i - \beta_{jk}\right)\right]}{1 + \exp\left[\alpha_j\left(\theta_i - \beta_{jk}\right)\right]}. \tag{2.10}$$

Therefore, each item holds a global discrimination parameter α_j (i.e., common to all response categories) and a set of location parameters β_{jk} for each response category. Since $P_{j0}^*(\theta_i, \pmb{p}_j) = 1$ regardless of the item parameters and ability level, the location parameter β_{j0} is actually not defined, so only location parameters β_{j1} to β_{j,K_j} are introduced, which also removes potential identification problems. Note also that by definition of the cumulative probabilities (2.8), the location parameters β_{jk}

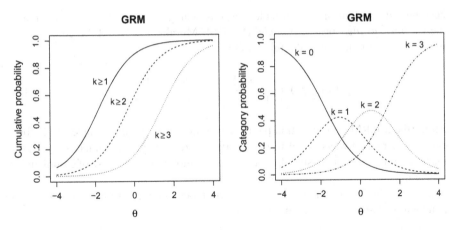

Fig. 2.2 Cumulative probabilities (left) and response category probabilities (right) of an artificial four-response item under GRM, with parameters $(\alpha_j, \beta_{j1}, \beta_{j2}, \beta_{j3}) = (1.2, -1.8, 0.3, 1.4)$

must increase with k in order to ensure the decreasing monotonicity of cumulative probabilities with k. Eventually, it is straightforward to notice that if $K_j = 1$ (i.e., in the case of only two response categories) then the GRM reduces to the 2PL model (2.5).

Figure 2.2 depicts both the cumulative probabilities (left panel) and the corresponding response category probabilities (right panel) for a hypothetical item with four responses under the GRM. The parameters chosen were as follows: $(\alpha_j, \beta_{j1}, \beta_{j2}, \beta_{j3}) = (1.2, -1.8, 0.3, 1.4)$. The cumulative probabilities have a logistic shape with the slope fixed by the α_j parameter (1.2 in this example) and the respective locations are set up by the β_{jk} parameters. Note that the cumulative probability for response category zero is not represented since it is equal to one for any ability level. The right panel represents typical response category probabilities for this type of polytomously scored item. Response category 0 has a decreasing probability since $P_{j0}(\theta_i, \boldsymbol{p}_j) = 1 - P_{j1}^*(\theta_i, \boldsymbol{p}_j)$, while response category 3 has an increasing probability since $P_{j4}(\theta_i, \boldsymbol{p}_j) = P_{j4}^*(\theta_i, \boldsymbol{p}_j)$ in this four-response case. For response categories 2 and 3, the corresponding probabilities are the differences between two successive cumulative probabilities from the left panel and display typical bell curves.

The GRM can be useful for ordered response categories (which justifies the use of cumulative probability functions) and can handle items with a different number of responses. However, some tests make use of the same set of categories for all items, for instance Likert scales with exactly the same number of points and responses. In this case, all items share the same number of location parameters β_{jk} (i.e., $K_j = K$ for all items) and it is possible to model them as being the sum of two components, one that is specific to the item (i.e., β_j) and another one that is specific to the response category (i.e., γ_k). In other words, the response category effect γ_k will be identical throughout all test items and the item-specific location effect β_j will become independent of the particular response category. This model is known as the *modified graded response model* (MGRM; Muraki, 1990) and the cumulative probability (2.8) then takes the following form:

$$P_{jk}^*(\theta_i, \boldsymbol{p}_j) = \frac{\exp\left[\alpha_j\left(\theta_i - \beta_j + \gamma_k\right)\right]}{1 + \exp\left[\alpha_j\left(\theta_i - \beta_j + \gamma_k\right)\right]}. \tag{2.11}$$

As for the MGRM, only location parameters γ_1 to γ_K are required since (2.11) for category zero is always equal to one. Furthermore, category parameters γ_k must increase along with k to ensure the corresponding monotonic decrease of cumulative probabilities.

In sum, both aforementioned difference models have various sets of item parameters, which can be summarized as follows:

$$\boldsymbol{p}_j = \begin{cases} (\alpha_j, \beta_{j1}, \ldots, \beta_{j,K_j}) & \text{for GRM} \\ (\alpha_j, \beta_j, \gamma_1, \ldots, \gamma_K) & \text{for MGRM} \end{cases} \tag{2.12}$$

This item parameter structure will be considered throughout the book, among others to accurately set up the item banks for adaptive and multistage testing.

2.2.2.2   Divide-by-Total Models

The second main class of polytomous IRT models are referred to as divide-by-total models, as their response category probabilities are defined as the ratio between some (category-related) mathematical functions and the sum of all these functions (across all response categories):

$$P_{jk}^*(\theta_i, \boldsymbol{p}_j) = \frac{\Gamma_{jk}(\theta_i, \boldsymbol{p}_j)}{\sum_{t=0}^{K_j} \Gamma_{jt}(\theta_i, \boldsymbol{p}_j)}. \tag{2.13}$$

One of the most popular divide-by-total models is the *partial credit model* (PCM; Masters, 1982). It extends the Rasch model (2.4) to the case of polytomously scored items with the difficulty level replaced by a set of threshold parameters δ_{jk}. Basically, it was developed for analyzing items for which partial credit can be awarded as the test taker achieves progressive levels of mastery. Using the notations by Embretson and Reise (2000), the PCM has the following response category probability characterization:

$$P_{jk}(\theta_i, \boldsymbol{p}_j) = \frac{\exp \sum_{t=0}^{k} (\theta_i - \delta_{jt})}{\sum_{r=0}^{K_j} \exp \sum_{t=0}^{r} (\theta_i - \delta_{jt})}, \tag{2.14}$$

i.e., $\Gamma_{jk}(\theta_i, \boldsymbol{p}_j) = \exp \sum_{t=0}^{k} (\theta_i - \delta_{jt})$. This model is over-parametrized and some identification constraints must be imposed onto the threshold parameters. One possibility is to constrain the δ_{jk} parameters to sum to zero. Another option (which will be retained in this book and is in line with Embretson and Reise (2000)) is to set $\sum_{t=0}^{0}(\theta_i - \delta_{jt}) = 0$, that is, the numerator of $P_{j0}(\theta_i, \boldsymbol{p}_j)$ is equal to one. Said differently, the threshold parameters of interest are $(\delta_{j1}, \ldots, \delta_{j,K_j})$. As will be illustrated later, they correspond to the ability levels for which the response category probabilities intersect. For instance, threshold δ_{j1} is the ability level for which response probabilities P_{j0} and P_{j1} cross, or equivalently, the ability level that makes both response categories 0 and 1 equally probable.

Two suggested extensions of the PCM are briefly outlined. The first one consists of adding an item-dependent discrimination parameter α_j for each item, that is (similar to the GRM) independent of the response categories. This yields the *generalized partial credit model* (GPCM; Muraki, 1992) and is written as:

$$P_{jk}(\theta_i, \boldsymbol{p}_j) = \frac{\exp \sum_{t=0}^{k} \alpha_j (\theta_i - \delta_{jt})}{\sum_{r=0}^{K_j} \exp \sum_{t=0}^{r} \alpha_j (\theta_i - \delta_{jt})}. \tag{2.15}$$

The identification constraint for the GPCM is the same as for the PCM, that is, $\sum_{t=0}^{0}(\theta_i - \delta_{jt}) = 0$, so that the numerator of $P_{j0}(\theta_i, \boldsymbol{p}_i)$ is equal to one. In other words, parameter α_j represents the "slope" parameter and impacts the respective slopes of the response category probabilities, while threshold parameters $(\delta_{j1}, \ldots, \delta_{j,K_j})$ play the same role as for the PCM.

Both models are illustrated in Fig. 2.3, for two artificial five-response items with the same threshold parameters. For the PCM (left panel), these thresholds are $(\delta_{j1}, \delta_{j2}, \delta_{j3}, \delta_{j4}) = (-1.5, -0.8, 0.5, 1.2)$. For the GPCM (right panel), the slope parameter α_j is fixed to 1.6. The two extreme response categories (0 and 4 here) have decreasing and increasing probabilities respectively, while all intermediate response categories exhibit bell-curve probability shapes. The four threshold parameter values can also be directly pointed out on the graph as the corresponding intersection points of two successive probability curves. Those points are identical for both PCM and GPCM items. The only difference between the models is that with the GPCM, the increased discrimination parameter leads to steeper probability curves (for extreme response categories) and more acute bell curves (for intermediate response categories).

The second extension of the PCM is actually similar to the extension of the GRM (2.10) to the MGRM (2.11). When all the test items exhibit exactly the same K response categories, the threshold parameters δ_{jk} can actually be split into an item-specific location parameter δ_j (common to all response categories) and a response threshold parameter λ_k (common to all items). This yields the so-called *rating scale model* (RSM; Andrich, 1978):

$$P_{jk}(\theta_i, \boldsymbol{p}_j) = \frac{\exp \sum_{t=0}^{k}(\theta_i - [\delta_j + \lambda_t])}{\sum_{r=0}^{K} \exp \sum_{t=0}^{r}(\theta_i - [\delta_j + \lambda_t])}. \tag{2.16}$$

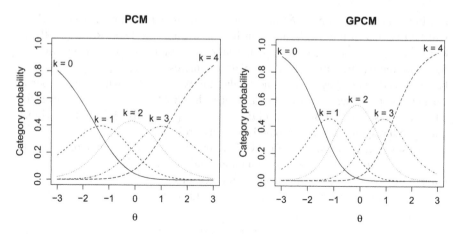

Fig. 2.3 Response category probabilities of an artificial five-response item under PCM (left) and GPCM (right). The threshold parameters are $(\delta_{j1}, \delta_{j2}, \delta_{j3}, \delta_{j4}) = (-1.5, -0.8, 0.5, 1.2)$ for both panels, and the slope parameter of the GPCM is $\alpha_j = 1.6$

For the RSM, one of the threshold categories must be fixed to avoid identification issues. This can be done, for instance, by imposing $\sum_{t=0}^{0}(\theta_i - [\delta_j + \lambda_t]) = 0$, that is, setting the numerator of $P_{j0}(\theta_i, \boldsymbol{p}_j)$ to be equal to one. In other words, only threshold parameters λ_1 to λ_K must be estimated (together with the item-specific location parameters δ_j).

The last polytomous IRT model to be presented here is the *nominal response model* (NRM; Bock, 1972). Instead of setting the Γ_{jk} function in the numerator of (2.13) as cumulative sums of exponentials, it provides a more general exponential form by setting item-specific and category-specific intercept and slope parameters c_{jk} and α_{jk}. The general NRM takes the following form:

$$P_{jk}(\theta_i, \boldsymbol{p}_j) = \frac{\exp(\alpha_{jk}\theta_i + c_{jk})}{\sum_{r=0}^{K_j} \exp(\alpha_{jr}\theta_i + c_{jr})}. \qquad (2.17)$$

This model is actually a generalization of the PCM and related extended models, which can be recovered from (2.17) by appropriately constraining the α_{jk} and c_{jk} parameters (Ostini & Nering, 2006). Usually, constraints must be set on the item parameters for model identification and the chosen set of constraints is to impose $\alpha_{j0} = c_{j0} = 0$.

Two artificial four-response items under NRM are illustrated in Fig. 2.4. Both share the same c_{jk} parameters but differ in terms of α_{jk} parameters. More precisely, $(c_{j1}, c_{j2}, c_{j3}) = (1.5, 2, 0.8)$ for both items; $(\alpha_{j1}, \alpha_{j2}, \alpha_{j3}) = (1, 2, 3)$ for item displayed on left panel; and $(\alpha_{j1}, \alpha_{j2}, \alpha_{j3}) = (1, 1.5, 2)$ for item displayed on right panel. The parameters for the left-panel item were chosen such that they actually correspond to an item calibrated under PCM with thresholds $(1.5, -0.5, 1.2)$, which explains why response category probabilities display similar curves as in Fig. 2.3.

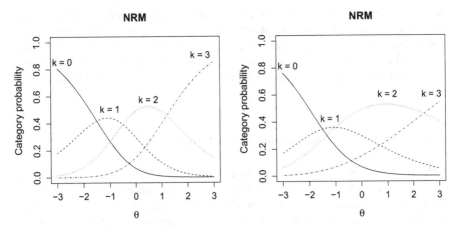

Fig. 2.4 Response category probabilities of two artificial four-response items under NRM with $(c_{j1}, c_{j2}, c_{j3}) = (1.5, 2, 0.8)$ and different α_{jk} parameters. Left panel: $(\alpha_{j1}, \alpha_{j2}, \alpha_{j3}) = (1, 2, 3)$. Right panel: $(\alpha_{j1}, \alpha_{j2}, \alpha_{j3}) = (1, 1.5, 2)$

The right panel, however, has different α_{jk} parameters and this directly impacts the shape of the response probabilities without affecting their relative ordering. This example clearly highlights the NRM's enormous flexibility in representing many types of response probability curves.

In summary, the divide-by-total models take the general form (2.13) with some specific sets of item parameters, depending on the model structure, which are summarized below:

$$
p_j = \begin{cases}
(\delta_{j1}, \ldots, \delta_{j,K_j}) & \text{for PCM} \\
(\alpha_j, \delta_{j1}, \ldots, \delta_{j,K_j}) & \text{for GPCM} \\
(\delta_j, \lambda_1, \ldots, \lambda_K) & \text{for RSM} \\
(\alpha_{j1}, c_{j1}, \ldots, \alpha_{j,K_j}, c_{j,K_j}) & \text{for NRM.}
\end{cases}
\tag{2.18}
$$

The ordering of the p_j parameters in (2.18) is arbitrary, but will be of primary importance for the design of the item banks in forthcoming chapters. Thus, we will use this particular ordering in the rest of the book.

2.2.3 Multidimensional IRT Models

Unidimensional models are very useful and powerful when all items are designed to target and evaluate a single, common latent trait of the test takers. However, in practice one might be interested in more than one such ability level. For instance, a test could include items that are all unidimensional, but target different abilities (for instance, large-scale assessments such as the PISA survey administer items that target proficiency in either mathematics, reading or science). It may also be the case that items measure more than one dimension, for instance, mathematics items that would also involve skills in reading comprehension. Very often those latent traits are correlated and it is therefore impossible to run independent unidimensional analyses for each latent dimension. Therefore, multidimensional IRT (or MIRT) models are required.

2.2.3.1 MIRT Models for Dichotomous Responses

MIRT models consider a vector of m latent ability levels $\boldsymbol{\theta}_i = (\theta_{i1}, \ldots, \theta_{im})$ instead of a single level θ_i. Most unidimensional IRT models have a multidimensional extension. For dichotomous responses, the *multidimensional 2PL model* (M2PL; McKinley & Reckase, 1982) holds a single location parameter but has a set of discrimination parameters (one parameter per latent dimension). The natural extension of the 3PL model to the *multidimensional 3PL* model (M3PL) involves including a lower asymptote, in a similar manner as extending the 2PL to the 3PL (Reckase, 2009).

The Rasch model can also be extended to this multidimensional framework, but not in such a straightforward manner as for the 2PL and 3PL models. Indeed, setting all discrimination parameters to a common constant actually leads back to the simple unidimensional 1PL model. For dichotomous responses, the *multidimensional Rasch model* (M1PL) takes the same mathematical form as the M2PL model, but now the slope parameters are not estimated from the data but fixed by the user (Adams, Wilson, & Wang, 1997). In other words, they act as factor loadings that describe to what extent each item contributes to each dimension. If all items contribute to a single dimension, the model is said to be a *between-item multidimensional* model. On the other hand, if the items contribute to more than one dimension, the model is said to be *within-item multidimensional*.

Finally, other multidimensional models for dichotomous responses were introduced. The former extensions (M1PL, M2PL and M3PL) belong to the class of so-called *compensatory MIRT models*. This class of models has an interesting property: any decrease in one latent dimension can be compensated by an appropriate increase in other dimensions such that the response probability remains constant. This is due to the fact that both latent traits and item parameters are linked through a linear combination that allows such a compensation. However, other MIRT models were introduced with the underlying framework of multiplicative effects of the latent dimensions, thus enabling this type of compensation in ability levels. These models are referred to as *non-compensatory* (or *partially compensatory*) *MIRT models*. For instance, the 3PL model has a compensatory MIRT version but also a non-compensatory extension, proposed by Sympson (1978). Further simplifications of this non-compensatory 3PL model yield non-compensatory versions of simpler IRT models, such as the non-compensatory multidimensional Rasch model (Maris, 1995). The interested reader can find more models and further details in Reckase (2009).

2.2.3.2 MIRT Models for Polytomous Responses

Several polytomous IRT models also have a multidimensional extension. The generalized partial credit model (2.15) was extended in the multidimensional context by Yao and Schwarz (2006). Similar to the extension of the 1PL model to the multidimensional Rasch model, Kelderman and Rijkes (1994) introduced a generalization of the partial credit model in a multidimensional context with (pre-defined) scoring weights for each response category on each dimension. These scoring weights must be pre-specified by the user, in a similar fashion as the multidimensional Rasch model. Finally, the graded response model (2.10) also has a multivariate extension that was introduced by Muraki and Carlson (1993). This extension requires computing the cumulative probabilities using a probit function. Note that this model was actually introduced by Samejima (1974) but with a slightly different approach. We refer the interested reader to Reckase (2009) for further details.

2.2.4 Other IRT Models

There are other types of IRT models for analyzing more complex data or for relaxing specific assumptions of traditional IRT models. These models are current research topics and none, or very few of them have been considered in an adaptive testing context. They are nevertheless briefly mentioned in this section for completeness.

2.2.4.1 Testlet IRT Models

Testlets are subsets of items that can be seen as a single unit and are therefore administered altogether (Wainer et al., 2007). They usually relate to a common context or stimulus, and it is, therefore, not rare that the related item responses are correlated to each other. In this regard, usual IRT models are inefficient at characterizing the relationships between items within a testlet. Moreover, they rely on the assumption of local item independence, which is clearly violated when testlets are present.

To overcome these issues, two approaches were proposed. In the first approach, the testlet can be seen as a kind of "super-item" whose score is actually the sum of all item scores of the testlet. Then, an appropriate polytomous IRT model can be applied to characterize each "testlet score", for example the nominal response model (2.17). In the second approach, traditional IRT models are modified to include a testlet effect, that is, an interaction effect between the person and the item nested into the testlet in question (Wainer, Bradlow, & Du, 2000). Most often, Bayesian approaches are required to estimate the model parameters (see Wainer, 2000, for further details and useful references).

2.2.4.2   Response-Time IRT Models

Most IRT models use the item responses as the only source of information for estimating the test takers' latent abilities (together with appropriate item characteristics). However, advances in both software development and processing power now allow researchers and users to record item response time, i.e., the time from the moment the item is presented to the test taker to the moment he/she responds. Obviously this requires the test to be administered via computer (but not necessarily adaptively).

Related to this advancement in data collection, appropriate IRT models were developed to estimate ability levels by considering the item responses and the response time conjointly. The inclusion of response time as an additional observation permits an increase in the precision of ability level estimates. One of the first attempts was proposed by Roskam (1987), who introduced a *speed* parameter into the Rasch model. A simplification of this model was introduced by Verhelst, Verstralen, and Jansen (1997), in which the person-item speed parameter is replaced by a simple person speed parameter. Furthermore, Wang and Hanson (2005) extended the 3PL model to incorporate response time data.

Other response-time IRT models and related algorithms were proposed, mostly based on Bayesian estimation procedures. This is still an area of active research. We refer the interested reader to references such as Klein Entink, Fox, and van der Linden (2009), Roskam (1997) and van der Linden, Klein Entink, and Fox (2010) for deeper insight into this appealing extension of IRT models.

2.3   Parameter Estimation

In the previous section, various IRT models were introduced according to the type of item response (dichotomous versus polytomous) and the number of dimensions (unidimensional versus multidimensional). These models hold parameters of two types: item parameters p_j and person parameters θ_i. Once data are collected from administering those items to a sufficiently large set of test takers, both item and person parameters can be estimated with accurate estimation procedures. This is often performed in two steps. First, item parameters are estimated independently of the ability levels: this is the *model calibration* step. Second, keeping the item parameters as fixed to their estimated values, the ability levels are then estimated from the response patterns of the test takers.

In adaptive testing situations, the calibration of the item bank (i.e., the estimation of all item parameters from all test items that constitute the bank of items) is assumed to be performed using a calibration sample or by some appropriate pre-testing (or field-testing) of the items of interest. The primary interest is the individual estimation of ability levels with some pre-specified adaptive item administration scenario. As a comprehensive introduction to IRT, both parts (model calibration and ability estimation) are discussed hereafter; however, emphasis and focus are put on the latter. For the sake of simplicity, only unidimensional IRT models under the general polytomously scored responses, will be considered.

2.3.1 Model Calibration

Model calibration consists of maximizing the full model likelihood $L(\theta, p)$ with respect to all item parameters $p = (p_1, \ldots, p_J)$. Because of the dependency of this full likelihood on the person parameters $\theta = (\theta_1, \ldots, \theta_I)$, direct maximization is technically impossible. Thus, specific techniques were developed to overcome this computational flaw. The three most-known estimation methods are joint maximum likelihood, conditional maximum likelihood and marginal maximum likelihood.

As the name indicates, *joint maximum likelihood* (JML) aims to conjointly maximize the full data likelihood with respect to both the person parameters θ and the item parameters p. To overcome the computational complexity of this task, an iterative approach is often applied, first by fixing the item parameters to constant values and maximizing $L(\theta, p)$ with respect to θ, then by fixing θ to the current estimated values and maximizing $L(\theta, p)$ with respect to p. This process is repeated until convergence of the model parameters. JML is easy to implement and computationally efficient. However, it also has several drawbacks: notably, it cannot be applied when the responses from one test taker, or to one item, are all correct or all incorrect, and the JML item estimates are known to be biased and inconsistent. Further information can be found in Embretson and Reise (2000) and Holland (1990).

Conditional maximum likelihood (CML) relies on a very particular property of some specific IRT models. It is indeed sometimes possible to derive sufficient statistics to estimate the latent ability levels. For instance, under the Rasch model, the total test score $r_i = \sum_{j=1}^{J} X_{ij}$ is a sufficient statistic for test taker's ability θ_i. In other words, two test takers with the same test score will obtain the same ability estimate (independently of which items were answered correctly or incorrectly). Based on this idea, it is then possible to rewrite the full likelihood by replacing the unknown ability levels θ with some known sufficient statistic. This allows for the estimation of the item parameters without requiring the estimation of person parameters, which is a major advantage to the JML approach. Moreover, it has been shown that the CML estimates are efficient, asymptotically normally distributed (Andersen, 1970), and conditions for existence and uniqueness were also derived (Fischer, 1981). However, CML has several drawbacks: it is only applicable for the broad class of Rasch models, it is inefficient with respect to constant response patterns (i.e. only correct or only incorrect dichotomous responses, or only first or only last response categories for polytomous items), and it can become very computationally intensive with long tests.

Finally, *marginal maximum likelihood* (MML) also aims to remove the person parameters from the maximization of the full likelihood, but instead of substituting them by some sufficient statistics (as for CML), the MML approach aims to integrate out the ability levels from the maximization process (Bock & Lieberman, 1970). More precisely, given an acceptable prior distribution for each person parameter, the marginal likelihood (that is, the expectation) of a response pattern can be written down, and the item parameters can be estimated by maximizing the full likelihood calculated as the product of all marginal pattern likelihoods. The prior distribution of ability levels is often set as normal, but other non-normal distributions can be specified (Mislevy, 1984). Despite its conceptual simplicity, computing the response pattern probabilities as expectations was conceptually complex, until Bock and Aitkin proposed the EM algorithm (Expectation-Maximization; Bock & Aitkin, 1981).

The MML approach has many advantages. It can easily handle perfect response patterns, unlike CML an JML, and is applicable to many types of IRT models, including polytomous and multidimensional models. Moreover, MML estimates can be used for likelihood-ratio tests to compare the fit of two nested models, for instance 1PL versus 2PL (Bock & Aitkin, 1981; Mislevy, 1986). Finally, despite its computational complexity, efficient EM algorithms are now commonly implemented in statistical software for IRT model calibration. The main drawback of MML is that it requires prior ability distributions to be set appropriately. However, it is not clear to what extent the misspecification of these distributions through inadequate priors will affect the calibration process (Embretson & Reise, 2000). For this reason, most often the prior is simply set as normal, though other non-normal distributions or empirical data-driven distributions can be implemented instead.

2.3.2 Ability Estimation

Once the item parameters are estimated from an appropriate calibration method, they are commonly set as fixed values and ability levels can then be estimated. The most well-known ability estimators are *maximum likelihood* (Lord, 1980), *maximum a posteriori*, also known as *Bayes modal* (Birnbaum, 1969); *expected a posteriori* (Bock & Mislevy, 1982), *weighted likelihood* (Warm, 1989), and *robust estimator* (Mislevy & Bock, 1982; Schuster & Yuan, 2011). They are briefly described below in the context of unidimensional polytomous IRT models (extensions to multidimensional IRT models are straightforward). Note that to simplify the notations, the test taker subscript i will be removed from this section, so that one focuses on the estimation of one's ability level θ with appropriate techniques.

The maximum likelihood estimator (MLE) of ability is the value $\hat{\theta}_{ML}$ which is obtained by maximizing the *likelihood function*:

$$L(\theta) = \prod_{j=1}^{J} \prod_{k=0}^{K_j} P_{jk}(\theta, \boldsymbol{p}_j)^{Y_{jk}}, \qquad (2.19)$$

(where Y_{jk} equals one if response category k for item j was chosen and zero otherwise), or its logarithm, the so-called *log-likelihood function*:

$$l(\theta) = \sum_{j=1}^{J} \sum_{k=0}^{K_j} Y_{jk} \log P_{jk}(\theta, \boldsymbol{p}_j), \qquad (2.20)$$

with respect to θ. Note that all other parameters are fixed so the (log-)likelihood function only depends on θ. The MLE has nice statistical properties: it is asymptotically unbiased, consistent, and its standard error is simply connected to the test information function (described later). Its main drawback, however, is that it does not take a finite value when the responses all come from the first or all from the last category (with dichotomous items, this corresponds to a constant pattern of only correct or only incorrect responses). This can be problematic with short tests or at early stages of adaptive testing.

Bayes modal (BME) or maximum a posteriori (MAP) estimator $\hat{\theta}_{BM}$ is obtained by maximizing the *posterior distribution* $g(\theta)$ of ability, that is, the product of the likelihood function (2.19) and the *prior distribution* of ability, say $f(\theta)$. In other words, $\hat{\theta}_{BM}$ is obtained by maximizing

$$g(\theta) = f(\theta)\,L(\theta) = f(\theta) \prod_{j=1}^{J} \prod_{k=0}^{K_j} P_{jk}(\theta,\boldsymbol{p}_j)^{Y_{jk}}, \qquad (2.21)$$

or equivalently by maximizing the log-posterior distribution:

$$\log g(\theta) = \log f(\theta) + l(\theta) = \log f(\theta) + \sum_{j=1}^{J} \sum_{k=0}^{K_j} Y_{jk} \log P_{jk}(\theta,\boldsymbol{p}_j). \qquad (2.22)$$

The prior distribution $f(\theta)$ must reflect some prior belief about the distribution of the latent trait of interest among the target population. Some common choices of parametric prior distributions are the standard normal density or the uniform distribution onto some predefined interval (for instance $[-4; 4]$). Another possibility is the so-called non-informative *Jeffreys' prior* (Jeffreys, 1939, 1946). It is defined as follows: $f(\theta) = \sqrt{I(\theta)}$, where $I(\theta)$ is the *test information function* (TIF):

$$I(\theta) = \sum_{j=1}^{J} I_j(\theta), \qquad (2.23)$$

and $I_j(\theta)$ is the *item information function* (IIF; Samejima, 1969, 1994):

$$I_j(\theta) = -E\left(\frac{d^2}{d\theta^2} l_j(\theta)\right) = -E\left(\frac{d^2}{d\theta^2} \prod_{k=0}^{K_j} Y_{jk} \log P_{jk}(\theta,\boldsymbol{p}_j)\right). \qquad (2.24)$$

Note that (2.24) can be written in a simpler form by making use of the first $P'_{jk}(\theta,\boldsymbol{p}_j)$ and second $P''_{jk}(\theta,\boldsymbol{p}_j)$ derivatives of the probability function $P_{jk}(\theta,\boldsymbol{p}_j)$ with respect to θ (Magis, 2015b).

Instead of maximizing the posterior distribution, i.e. determining the mode of this distribution, one can compute its expected value. This is the aim of the expected a posteriori estimator (EAP):

$$\hat{\theta}_{EAP} = \frac{\int_{-\infty}^{+\infty} \theta \, g(\theta) \, d\theta}{\int_{-\infty}^{+\infty} g(\theta) \, d\theta}. \tag{2.25}$$

Both BME and EAP rely on the accurate selection of a prior distribution, which can be difficult to determine from the target population. Aside from that, both Bayesian estimators have reduced variability with respect to MLE (due to the well-known shrinkage effect that shrinks the ability estimates towards the mean of the prior distribution) and are always finite (provided the prior distribution converges towards zero at the extremes of the ability scale).

The weighted likelihood estimator (WLE) $\hat{\theta}_{WL}$, originally proposed by Warm (1989) for dichotomous scored items and later extended by Samejima (1998) for polytomous items, was introduced to reduce the bias of the MLE when asymptotic conditions are not met (i.e., with short tests or at early stages of adaptive testing). It is determined by solving the following equation:

$$\frac{J(\theta)}{2\,I(\theta)} + \frac{d}{d\theta} l(\theta) = 0, \tag{2.26}$$

where $I(\theta)$ is the TIF (2.23) and

$$J(\theta) = \sum_{j=1}^{J} \sum_{k=0}^{K_j} \frac{P'_{jk}(\theta, \mathbf{p}_j) \, P''_{jk}(\theta, \mathbf{p}_j)}{P_{jk}(\theta, \mathbf{p}_j)}. \tag{2.27}$$

Equation (2.26) can be seen as the first derivative of a weighted version of the log-likelihood function (2.20), i.e., $\omega(\theta) \, L(\theta)$ where the weight function $\omega(\theta)$ is such that

$$\frac{d\omega(\theta)}{d\theta} = \frac{J(\theta)}{2\,I(\theta)}. \tag{2.28}$$

Warm (1989) established that the WLE is identical to the BME with Jeffreys' prior distribution. This results was extended to the classes of difference and divide-by-total polytomous IRT models (Magis, 2015c).

Finally, Mislevy and Bock (1982) proposed a different weighting system to cope with the problem of aberrant responses and their impact on ability estimation. Aberrant responses have various origins, such as cheating, guessing, inattention, tiredness etc. Such responses obviously affect the likelihood function and subsequently the estimated ability levels. The idea of robust estimation is to weight the likelihood function (more precisely its first derivative) in order to reduce the impact of aberrant responses onto the full likelihood and limit the corresponding bias. First set

$$l_j(\theta) = \sum_{k=0}^{K_j} Y_{jk} \, \log P_{jk}(\theta, \mathbf{p}_j), \tag{2.29}$$

so that the log-likelihood (2.20) can be simply written as $l(\theta) = \sum_{j=1}^{J} l_j(\theta)$. Then, the MLE is obtained by maximizing $l(\theta)$ with respect to θ, which is equivalent to solving

$$\frac{d\,l(\theta)}{d\theta} = \sum_{j=1}^{J} \frac{d\,l_j(\theta)}{d\theta} = 0. \tag{2.30}$$

The robust estimator $\hat{\theta}_{ROB}$ is computed by solving a weighted version of (2.30):

$$\sum_{j=1}^{J} \omega_j(\theta) \frac{d\,l_j(\theta)}{d\theta} = 0. \tag{2.31}$$

The weight functions $\omega_j(\theta)$ must be appropriately specified so that abnormal responses (e.g., incorrect responses to easy items) have lower weight and hence reduced impact on the estimation process. Mislevy and Bock (1982) recommended using the so-called Tukey *biweight* (or *bisquare*) weight function (Mosteller & Tukey, 1977), while Schuster and Yuan (2011) advocated the use of Huber-type weight functions. More details can be found in Magis (2014) and Schuster and Yuan (2011).

2.3.3 Information and Precision

The previous section described the best known methods for estimating ability under a general (unidimensional) IRT framework. Although the point ability estimates are informative, it is also of primary importance to evaluate their variability through estimated standard errors of measurement (SEs). Small standard errors indicate small levels of variability of the ability estimator and hence high levels of precision in this point estimate. Using asymptotic considerations, easy-to-use SE formulas can be derived for each estimator.

First, the SE of the MLE is computed as the inverse square root of the TIF (2.23) at the point MLE estimate:

$$SE_{ML}(\hat{\theta}_{ML}) = \frac{1}{\sqrt{I(\hat{\theta}_{ML})}}, \tag{2.32}$$

Thus, the larger the TIF, the smaller the SE and the greater the precision of the MLE. Note that since the TIF (2.23) is additive with respect to the number of items, longer tests usually lead to smaller SE values.

The BME estimator has an associated SE formula which incorporates the prior distribution as follows (see Wainer, 2000):

$$SE_{BM}(\hat{\theta}_{BM}) = \frac{1}{\sqrt{I(\hat{\theta}_{BM}) - \left(\frac{d^2 f(\theta)}{d\theta^2}\right)_{\theta = \hat{\theta}_{BM}}}}. \tag{2.33}$$

Because the prior distribution is assumed to be convex at the ability estimate, the second derivative in the denominator of (2.33) is positive. For instance, using the normal density $N(\mu, \sigma^2)$ as the prior distribution, the SE (2.33) becomes

$$SE_{BM}(\hat{\theta}_{BM}) = \frac{1}{\sqrt{\frac{1}{\sigma^2} + I(\hat{\theta}_{BM})}}. \tag{2.34}$$

Equations (2.33) and (2.34) indicate that the SE of the BME is usually smaller than that of the MLE, because of the inclusion of the prior distribution in its computation. This is the translation of the well-known shrinkage effect for Bayesian estimators of ability (Lord, 1986). Note that for dichotomous IRT models, Magis (2016) proposed a different SE formula based on a broader class of ability estimators that include the BME.

For the EAP, the asymptotic SE is computed as the square root of the variance of the posterior distribution (Bock & Mislevy, 1982):

$$SE_{EAP}(\hat{\theta}_{EAP}) = \sqrt{\frac{\int_{-\infty}^{+\infty} (\theta - \hat{\theta}_{EAP})^2 \, g(\theta) \, d\theta}{\int_{-\infty}^{+\infty} g(\theta) \, d\theta}}. \tag{2.35}$$

With respect to the WLE, Warm (1989) established that the asymptotic distribution of the WLE is identical to that of the MLE, and therefore suggested using the same SE formula as the MLE, i.e. $SE_{WL}(\theta) = SE_{ML}(\theta)$, but by plugging-in the WLE estimate $\hat{\theta}_{WL}$ instead of the MLE estimate:

$$SE_{WL}(\hat{\theta}_{WL}) = \frac{1}{\sqrt{I(\hat{\theta}_{WL})}}. \tag{2.36}$$

However, alternative formulas were developed, for instance by Magis (2016).

Finally, the robust estimator has its own SE formula in case of dichotomous IRT models (Magis, 2014):

$$SE_{ROB}(\hat{\theta}_{ROB}) = \frac{\sqrt{\sum_{j=1}^{J} \omega_j(\hat{\theta}_{ROB})^2 I_j(\hat{\theta}_{ROB})}}{\sum_{j=1}^{J} \omega_j(\hat{\theta}_{ROB}) I_j(\hat{\theta}_{ROB})}, \tag{2.37}$$

where $\omega_j(\theta)$ is the robust weight function (e.g., Tukey biweight or Huber-type) chosen to compute the robust estimate in (2.31).

2.4   Further Topics

This brief overview was a short introduction to IRT with emphasis on the main aspects that will be useful for adaptive and multistage testing applications. Several interesting and related topics were not covered here but constitute complimentary materials in the IRT framework. Some such topics are: (a) dimensionality assessment, (b) local item independence testing, (c) model fit and person fit analysis, (d) differential item functioning, and (e) linking and equating. All these topics are of primary importance for any IRT analysis or for assessing the calibration accuracy of an item bank with field-tested data. However, adaptive testing makes the strong assumption that the item bank under consideration was previously calibrated and checked for accurate dimensionality, model, item and person fit, and is free of item bias or differential item functioning. Therefore, only a brief overview of these concepts are presented here. A complete and concise overview of these topics can be found in DeMars (2010).

2.4.1   Dimensionality

Section 2.2 presented the main unidimensional and multidimensional IRT models. Choosing between the two classes of models is not an a priori option for the test developer or analyst: one has to know whether the test items target one or several latent dimensions. The number of latent dimensions one has to consider for proper analysis and model fit can be determined by examining the collected data.

There are several possible approaches to consider when focusing on dimensionality. One well-known approach involves computing the eigenvalues of the tetrachoric correlation matrix of item responses, and then plotting them in decreasing order, leading to the so-called *scree plot*. The number of latent dimensions can then be evaluated in various ways, for instance by determining the eigenvalues that are much larger than the remaining ones. If one eigenvalue is significantly larger than the others, then there is good evidence for the presence of one main latent dimension. Other rules based on the numeric values of the eigenvalues have also been developed (Reckase, 1979; Zwick & Velicer, 1986).

There are other commonly used approaches, such as the *test of essential unidimensionality* (Stout, 1987), also simply referred to as DIMTEST (Stout, 2005). Another is based on the *analysis of residuals from unidimensional model* (Finch & Habing, 2007), also known as the *NOHARM-based methods* (Fraser & McDonald, 2003), and associated indices such as the chi-square index (Gessaroli & De Champlain, 1996). Expanding on all available methods for dimensionality assessment is beyond the scope of this book, we therefore refer the interested reader to Hattie (1984) and Tate (2003) for detailed treatments of the many available methods.

2.4.2   Local Item Independence

If the assumption of unidimensionality cannot be established with any of the methods described in the previous section, then multidimensional IRT models can be considered instead. The assumption of local item independence, on the other hand, is more crucial in this case. As seen in Sect. 2.3, local independence is required for using traditional methods to calibrate the items and estimate ability levels.

The most commonly used approach to test for local (in)dependence is to use Yen's Q_3 index (Yen, 1984, 1993). This method checks all pairs of items for local dependence by computing the correlation between the residuals (i.e., expected probability minus observed response) of the two items. Under the assumption of local independence, these residuals should be uncorrelated, so large residual correlations (i.e., Q_3 statistics) are evidence of pairwise local dependence. Other such indices were proposed (Chen & Thissen, 1997) though Q_3 remains, apparently, the most popular one.

If some items exhibit local dependence, at least two options can be considered. The first option is to assume that these items belong to a common testlet and then use any of the IRT testlet models currently available (Sect. 2.2.4.1). The second is to modify the likelihood function by imposing some interaction between the parameters of the items that are locally dependent. One recent approach in this direction is the use of *copula functions* (e.g., Braeken, Tuerlinckx, & De Boeck, 2007).

2.4.3   Model and Person Fit

Once an IRT model is chosen and calibrated, it is vital to ensure that the model fits the data to an acceptable extent. Several approaches are available, most of them based on the idea of comparing the item characteristic curves of the fitted model to the observed proportion of correct responses (or the proportions of responses per category for polytomous IRT models). It was proposed that the fit can be visually assessed for each item or tested by using chi-square based statistics (Bock, 1972; Muraki & Bock, 2003; Yen, 1981). DeMars (2010) provides a good review of these issues and suggested solutions. Alternatively, evaluating the overall fit can be done for all items simultaneously, for instance by comparing the observed total score distribution to the expected, model-based score distribution (Swaminathan, Hambleton, & Rogers, 2007).

Additionally, once it is established that the calibrated IRT model fits the data, it is necessary to address the issue of *person fit* by checking that the estimated ability levels were not affected by aberrant response patterns. Many indices have been proposed to determine whether the observed responses globally match the expected responses probabilities arising from the calibrated IRT model. Some well known examples are the Infit (Wright & Masters, 1982) and Outfit (Wright & Stone,

1979) indexes, the standardized l_z index (Drasgow, Levine, & Williams, 1985) and its corrected version l_z^* (Snijders, 2001). A detailed review of person fit indices is provided by Karabatsos (2003).

2.4.4 Differential Item Functioning

It is important to ensure that the item properties remain the same across various groups of test takers. For instance, it may be the case that some items favor (e.g., are easier for) boys over girls (controlling for equal ability levels). This phenomenon is known as *differential item functioning* (DIF) and items exhibiting this phenomenon must be identified (and possibly removed) to ensure fair measurement.

There are many DIF detection methods, most of them for unidimensional dichotomous items. Some are based on test scores, such as Mantel-Haenszel method (Holland & Thayer, 1988) or logistic regression (Swaminathan & Rogers, 1990), and aim to test whether the interaction between group membership and the item response (controlling for test score) is statistically significant (which is interpreted as an evidence for DIF). Other methods are based on IRT models, for instance Lord's test (Lord, 1980) which compares the item parameters of the same model in different subgroups of test takers (after concurrent calibration) or Likelihood-ratio test (Thissen, Steinberg, & Wainer, 1988) which tests between two nested models (one allowing for different item parameters across groups and the other with identical parameters across groups). There are many other DIF methods, depending on the type of DIF effect to detect, the number of subgroups of test takers to consider, and the item types (e.g., dichotomous versus polytomous responses). We refer to Magis, Béland, Tuerlinckx, and De Boeck (2010), Millsap and Everson (1993), Osterlind and Everson (2009) and Penfield and Camilli (2007) for excellent reviews on this topic.

2.4.5 IRT Linking and Equating

Lastly, an important IRT topic for practical CAT and MST purposes is the concept of *equating*. Equating is the statistical process used to adjust scores on alternative test forms so that the scores can be used interchangeably. Equating adjusts for differences in difficulty among forms that are built to be similar in difficulty and content (it does not adjust for differences in content though). In most equating designs, the test forms to be equated are administered to different groups of people.

When one estimates the IRT item parameters, certain constraints are usually set to make IRT models identified (e.g., mean 0 standard deviation 1). If the groups have different abilities, the ability distribution for each group is still scaled to have a mean of 0 and a standard deviation of 1. Thus the restrictions have different effects and a transformation of IRT scales is needed.

In CAT and MST operational programs, the linking of item parameters is necessary to put the new items on the same scale as the items from the item bank. In operational programs, in which different items are used at each test administration, new items need to be included in the item pool. The test administration and the test specifications are standardized to ensure test comparability over administrations. In order to always have a sufficient number of new items in the item bank, one needs to administer the new items to a representative sample from the test taker population.

The linking of the item parameters for the new items in CAT or MST resembles the non-equivalent groups linking design. The new items administered to a new population need to be linked to the items from the bank taken by the old population. The operational items in a CAT or MST may function as the common items for this linking. In general, if the estimation of item parameters is carried out separately on two samples from different populations, then parameter estimates for the "common items" are available for examinees in the two groups. These estimates serve as the basis for the scale transformation ("mean-mean" methods or test characteristics methods).

If an IRT model fits some data set, then any linear transformation of the θ-scale also fits the set of data. Thus, a linear equation (finding the parameters A and B) can be used to convert IRT parameter estimates to the same scale:

$$\theta_{new} = A\theta_{old} + B. \tag{2.38}$$

The main goal of the IRT linking is to estimate the linking parameters, A and B and apply them to the transformation for the alignment of the new item parameters with the ones from the bank.

Two equating methods are commonly used in operational tests: the mean-mean method and the Stocking-Lord method. In the *mean-mean* method, the mean of the a-parameter (discrimination) estimates for the common items is used to estimate the slope of the linear transformation. Then, the mean of the b-parameter (difficulty) estimates of the common items is used to estimate the intercept of the linear transformation. In order to avoid problems with the mean-mean approach when various combinations of the item parameter estimates produce almost identical item characteristic curves over the range of ability at each test taker's score, characteristic curves transformation methods were proposed (Haebara, 1980; Stocking & Lord, 1983).

The *Stocking-Lord* method, finds the parameters A and B for the linear transformation of item parameters in one population that matches the test characteristic function of the common items in the reference population. With Stocking-Lord method, all the items may be estimated jointly, coding the items that a test taker did not take as "not administered," as these outcomes were unobserved and are (almost) missing by design (see the discussion about the local independence assumption above). This actually results in joint calibration. This method is the method of choice for CAT, because it is difficult to administer all the new items together to the same test takers. Moreover, concurrent calibration is also used for the calibration of the items in the bank in the first stages of building the item bank for adaptive tests, CAT or MST. This stage is often called "pre-equating".

Note finally that another linking method is *simultaneous linking*, which is conducted to adjust separately calibrated parameters and to correct for the measurement and sampling errors from the individual calibrations. It is used at regular intervals in the life of an assessment. This method is used only for operational programs that have an almost continuous administration mode, and where the risk of cumulative errors is high (Haberman & von Davier, 2014).

Part I
Item-Level Computerized Adaptive Testing

Chapter 3
An Overview of Computerized Adaptive Testing

In this chapter, we present a brief overview of computerized adaptive testing theory, including test design, test assembly, item bank, item selection, scoring and equating, content balance, item exposure and security. We also provide a summary of the IRT-based item selection process with a list of the commonly used item selection methods, as well as a brief outline of the tree-based adaptive testing.

3.1 Introduction and Background

Computerized adaptive testing (CAT) has been used to measure ability and achievement, and for measures of personality and attitudinal variables. The main purpose of CAT is to produce a test that is more efficient and accurate than a linear test. With its promising performance, CAT has become an important and popular area of research and practical implementations in the field of psychometrics and educational testing.

In order to increase the accuracy of computerized tests and the accuracy of the ability estimate, the error associated with an ability estimate needs to be reduced. The error associated with an ability estimate is a function of the information available for measurement. If information can be increased, then the magnitude of errors is correspondingly decreased. One way of increasing available information without increasing test length is tailored testing. CAT is a tailored test that adapts to each test taker's currently estimated ability level after each administered item, as will be outlined later.

Specifically, in a CAT framework, the tests involve an iterative administration of test items adapting to each test taker's estimated ability level, in order to produce more accurate ability estimates. In an ideal CAT, the items are selected and administered one by one in an optimal sequence for each test taker, such that each item selected is the most useful or informative item at the current step of the test. Also, the selection of the next item is conditional upon the previously administered

© Springer International Publishing AG 2017
D. Magis et al., *Computerized Adaptive and Multistage Testing with R*, Use R!,
https://doi.org/10.1007/978-3-319-69218-0_3

items, the responses from the test taker, and the provisional estimate of the test taker's ability level. CAT has many advantages with respect to linear testing; among others, it leads to shorter test for the test taker while providing the same or higher level of precision for ability estimation. It also produces ability estimates that are available directly after test administration for immediate feedback for the test takers (Magis & Raîche, 2012). It also reduces the risk of fraud or cheating, since each test taker has a different sequence of test items.

The volume of CAT literature is constantly increasing and the books by Wainer (2000), Mills, Potenza, Fremer, and Ward (2002), van der Linden and Glas (2010) and Yan, von Davier, and Lewis (2014) give thorough discussions of all important aspects of CAT theory, implementation, and applications. However, until recently there was a lack of flexible, open source software package for running CATs and performing intensive simulation studies within this framework. The R package **catR** (to be presented in the next chapter) was developed expressly to meet this goal.

3.2   CAT Basics

The elements of a CAT include an item bank containing pre-calibrated items, the process of selecting appropriate items, the process of estimating ability level after each item is administered, the stopping criteria and the final estimation and scoring. There are many practical issues to consider, such as item exposure and content balancing. In the case of a classification test, the final estimation will be a classification based on ability estimation.

Item response theory (IRT) methodologies are used in several processes of CAT, and focus on increasing the accuracy and efficiency of ability estimation. There are some basic IRT assumptions required for CAT applications. In this book it is assumed that the latent trait is unidimensional (i.e., *unidimensionality* assumption) and that item responses are conditionally independent given the latent trait or ability (i.e., *local item independence* assumption). These assumptions were briefly outlined in Sect. 2.1. They are quite common in adaptive testing (Weissman, 2014), though multidimensional CAT research programs have recently emerged (Reckase, 2009; Segall, 2010).

As Magis and Raîche (2012) described, any CAT process requires a calibrated item bank and can be schematically decomposed in four successive steps.

1. The first step is the *initial step* or *starting step* and consists of selecting one or several items to initialize the CAT.
2. The second step is the *test step*, in which the items are iteratively and optimally selected and administered, and the ability level estimate is re-computed after each item administration.
3. The third step is the *stopping step*, which defines the rules to make the adaptive item administration stop.
4. The *final step* yields the final estimation of ability and reporting options.

This CAT process is depicted in Fig. 3.1.

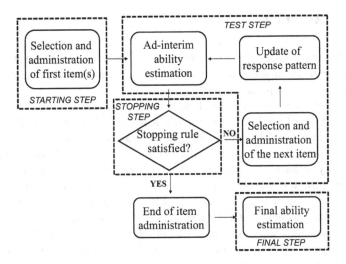

Fig. 3.1 Schematic illustration of a CAT process

3.3 Test Design and Implementation

Designing and implementing a CAT involves many considerations, such as the purpose of the test, that lead to different test designs, item bank design and maintenance, content balance and test assembly, scoring and equating, reliability and validity, test security and exposure control.

For the purpose of proficiency testing, CAT design is focused on the accuracy of ability measurement, that is, the estimation accuracy for a range of test takers' proficiency levels. IRT approaches are commonly used for test and item bank designs, test assembly, item calibration, item selection, and model parameter estimation. There are also IRT methods with a Bayesian approach for CAT (van der Linden, 1998a). For the purpose of classification testing, CAT design is focused on the accuracy of classification, that is, the accuracy of classifying test takers into appropriate groups. IRT approaches are also used for test and item bank designs, test assembly, item calibration, selection, and estimation for computerized mastery testing (CMT) for two categories (Lewis & Sheehan, 1990) and classification testing (Eggen, 2010; Glas & Vos, 2010; Smith & Lewis, 2014). There are other non-IRT based methodologies including a tree-based methodology for CAT (Yan, Lewis, & Stocking, 2004), see also Sect. 3.8.

The basic issues to consider when designing a CAT are: How long should a test be? How many items will be used? How will the first and subsequent items be selected? What are the mathematical models for item selection and ability estimation? What is the stopping criterion? How will the test be scored? What are the content balancing requirements? How to control for item exposure? All of these questions can be addressed and investigated in a straightforward manner with a reliable and easy to use software package like **catR**.

3.4   Test Assembly

Over the last two decades, many researchers have contributed to the development of test assembly methods for CAT. Currently, several automated test assembly (ATA) algorithms are used in practice.

Traditionally, item difficulty and item discrimination parameters in classical test theory (CTT) have been used for test assembly. With modern IRT, the test information function (2.23) became the primary statistic for test assembly. Because of the close connection between the TIF and the standard error (SE) of measurement for the ability or latent trait, the level of measurement error can be controlled by manipulating the test information function (Zheng, Wang, Culbertson, & Chang, 2014). Many researchers have proposed different approaches for test assembly, including Lord's method using the target test information curve (Lord, 1977) and Samejima's identical test information function for same ability levels (Samejima, 1977). The principle of matching TIFs is still a mainstream approach for test assembly methods (Zheng et al., 2014).

Commonly used automated test assembly (ATA) methods include linear programming methods that optimize an objective function over the binary space, subject to multiple constraints (van der Linden, 2008; van der Linden & Diao, 2014; van der Linden & Guo, 2005). The objective functions include the test information function, the deviation of the information of the assembled test from the target, and the differences among multiple parallel test forms. Heuristic methods use a sequence of local optimizations by each selecting a single item to add to the test (Ackerman, 1989; Lord, 1977). The objective functions include TIF, which is penalized by various "peripheral" constraints such as content coverage (Zheng et al., 2014).

Among these methods, the weighted deviation model (WDM; Swanson & Stocking, 1993) minimizes the weighted sum of (absolute) deviations, and the normalized weighted absolute deviation heuristic (NWADH; Luecht, 1998) normalizes each weighted deviation from constraint targets to a common scale. WDM and NWADH treat all constraints as targets and form the criterion as the weighted sum of (normalized) deviations from the targets. The maximum priority index (MPI; Cheng & Chang, 2009; Cheng, Chang, Douglas, & Guo, 2009) is multiplied by a factor computed from the number of remaining items permitted by each constraint.

3.5   The Item Bank

An item bank is a collection of items that can be administered to the test takers. It is the central tool for adaptive testing. For a CAT, the item bank is assumed to be calibrated prior to the start of the CAT process. That is, a large collection of items, which have been administered and calibrated from a pretest or field or pilot test, is available for item selection in adaptive testing.

To build a CAT item bank, test designers first design a blueprint that could support the assembly of a desired CAT (van der Linden, Veldkamp, & Reese, 2000). The blueprint should describe the item bank design based on test specifications, which are based on the purpose of the test, i.e., proficiency test or classification test, and should describe what kind of items would be needed for an optimal item bank for the test assembly problem at hand. Then, this blueprint can be applied to instruct and train item writers to write the items following the test specifications.

There are many methods for item bank design and maintenance, including integer programming models, heuristic methods, and Monte Carlo methods (Belov & Armstrong, 2009; van der Linden, 2000, 2005; van der Linden, Ariel, & Veldkamp, 2006; Veldkamp & van der Linden, 2010). An alternative approach would be to use the blueprint as a starting point for automated item generation instead of item writing (Irvine & Kyllonen, 2002; Veldkamp, 2014). One of the techniques used with this approach is item cloning, where families of clones are derived from a parent item, by varying those attributes that are assumed not to be related to the item's difficulty (Bejar et al., 2003; Geerlings, Glas, & van der Linden, 2011; Glas & van der Linden, 2003).

There are also methods to diagnose weaknesses of operational item banks by applying infeasibility analysis to identify what kinds of items are missing (Belov & Armstrong, 2009; Huitzing, Veldkamp, & Verschoor, 2005). Diagnostic information derived from these methods might be used to fine-tune the blueprint design (Veldkamp, 2014).

In summary, test designers need to obtain test specifications, create sufficient numbers of items in each content category, review the item quality with respect to specifications, review the items for fairness and perform initial pretesting or piloting of the newly written items. Once the pretesting is done, psychometricians and statisticians need to calibrate the items using classical test theory (CTT) or item response theory (IRT), and perform a statistical review for item quality using CTT or IRT criteria. Any items that do not meet quality specifications due to flaws or ambiguities, low correlation with total observed score, or with extremely low or high difficulty level, should be eliminated. Thus, only selected qualified items can be added to an item bank. The item bank will also be periodically evaluated/re-evaluated for its size, specifications, and content balance.

A frequently asked question is "how large does our item bank need to be?" The larger the item bank, the better it is for a CAT process using the bank. But in reality, it is not always possible to implement an extremely large item bank. Also, it is hard to say how large an item bank should be for optimal adaptive testing. A balanced item bank should contain items for the whole range of difficulty levels, from easy to difficult items. This enables accurate estimation of the whole range of ability levels, since easy items are most informative for low ability levels and difficult items are most informative for high ability levels. Including easy or difficult items allows accurate estimates of low or high ability test takers.

3.6   IRT-Based CAT

This section provides a brief overview of the most important aspects of IRT-based CAT by describing each of the four steps depicted in Fig. 3.1.

3.6.1   Initial Step

The goal of this first step is to select at least one item in the item bank and administer it to the test taker. Most often, a single item is selected, and without any prior information about the test taker's ability level, the first item is chosen as most informative around the prior population mean ability value (most often set to zero). But this common approach can be improved in many ways.

First, if some prior information about the test taker's ability level is available, it can be incorporated into the initial step. A few modifications include adjusting the prior mean ability level to the population mean ability level if it is higher or lower than zero, and selecting an initial item whose difficulty level is very close to the test taker's ability level if there is prior information about that level.

Second, the initial item can be selected with a different optimality criterion. Instead of determining the item that is most informative for the initial ability level, Urry (1970) proposed selecting an item whose difficulty level is closest to this initial ability level. This is referred to as the *bOpt* or *Urry*'s rule. Note that it is most suitable for dichotomous IRT models, since the difficulty level has no straightforward equivalent with polytomous IRT models (many threshold parameters are considered instead of a single difficulty level).

Third, it may be useful to select more than one item at the initial step. This is not common or clearly stated in the literature, nor available in typical CAT software. But this approach could be considered for selecting two or three items, each item referring to a different prior ability level in order to cover some range of abilities from the start of the CAT. In addition, one should avoid selecting the same initial item(s) for all test takers for content security reasons. An improvement on this approach is selecting, for each initial ability level, a small set of items (with respect to the chosen optimality rule) and to randomly sample one of these items for initial administration. This process maintains near optimality for the selected items while avoiding the administration of the same starting items (and hence limiting item over-exposure). This is known as the *randomesque* approach (Kingsbury & Zara, 1989).

Less optimal methods can also be considered, such as random selection of the starting item(s) or letting the test administrator decide which item(s) must be assigned to each test taker.

3.6.2   Test Step

The test step is where the adaptive part of the CAT process comes into play. After administering the starting item(s), the test step proceeds as follows:

1. Ability is estimated using the current set of item responses.
2. If the stopping rule is not satisfied:

    (a) the next item administered is selected from the eligible items according to the chosen item selection method (see Sect. 3.6.3);
    (b) the test taker's response is recorded and the response pattern is updated;
    (c) the process goes back to step 1.

3. If the stopping rule is satisfied, the test step stops.

Thus, the main technical aspects involved are the provisional estimation with the current response pattern and set of administered items, and the optimal selection of the next item. This section focuses on the ability estimation during the CAT process; details regarding item selection are provided in the next section.

As mentioned in Sect. 2.3.2, various ability estimators are available: maximum likelihood, weighted likelihood and Bayesian estimators are the most popular ones in CAT (Magis & Barrada, 2017; Magis & Raîche, 2012; van der Linden & Glas, 2010). Instead of the same ability estimator being used frequently throughout the test step, sometimes it is useful to employ a hybrid rule, i.e., starting with an estimator and then switching to another one after a certain number of administered items or when a particular condition is satisfied, such as, in the very first stages of a CAT, when only a few item responses are available. With only one or two such responses a *constant pattern* will likely be observed, that is, only correct or only incorrect responses (in case of dichotomously scored items), or only the first or only the last response categories (in case of polytomously scored items). In such a scenario, it is well-known that the ML estimator returns infinite estimates of ability (see also Sect. 2.3.2). This yields considerable instability in these first stages, by selecting items that are most informative for such extreme ability estimates and most probably not at all adequate with respect to the test taker's true ability level.

The most common approach to overcome this undesirable situation is to start with an ability estimator that always returns finite estimates, and then to switch to any desired ability estimator (including ML) when the response pattern is not constant anymore (or to continue the CAT with the same initial ability estimator). The BME or EAP with the prior distribution of ability that converges towards zero at the extremes of the ability scale, as well as the WLE for most common unidimensional IRT models, are adequate estimators for returning finite estimates of ability (Magis & Verhelst, 2017).

An alternative approach for dealing with constant patterns in the early stages of CAT (dichotomous items only) is to use heuristic adjustment methods as suggested by Dodd, De Ayala, and Koch (1995). They basically consist of starting with an initial guess of ability (usually zero) and then iteratively adjusting this initial guess with some stepsize increase or decrease (depending on whether the last response was correct or incorrect). For instance, if the current response pattern has only correct responses, then the current ability estimate will be increased by some value, forcing the next selected item to be more difficult and (hopefully) the next response will be incorrect, thus yielding a pattern that is no longer constant.

The value added or subtracted from the current ability estimate depends on the chosen heuristic stepsize adjustment method. Two approaches are commonly suggested: *fixed stepsize* adjustment, in which the step increase or decrease is constant and fixed by the test administrator; or *variable stepsize* adjustment, in which the current ability level is increased or decreased by half the distance between this current value and the largest (or smallest) difficulty level in the item bank.

According to Dodd et al. (1995), variable stepsize adjustment performs slightly better than fixed stepsize adjustment (considered fixed step values were 0.4 and 0.8). However, Magis (2015a) showed that making use of either BME or WLE at early stages of a CAT outperforms the overall CAT estimation process with respect to stepsize adjustment.

3.6.3 Item Selection Method

While item selection is a component of the test step, there are many methods for next item selection, thus it deserves accurate notation and a separate section for presentation and discussion.

Consider $t - 1$ items (with $t > 1$) having already been administered at the current stage of the CAT (either all during the starting step or some during the test step also). What we want to do at this point is select the t-th item to administer as the most optimal for CAT assessment. Let X_{t-1} be the current response pattern, made up of the first $t - 1$ responses, for the test taker of interest (in this section, person subscript i introduced in Chap. 2 is removed to shorten notation). The set S_t will denote the set of eligible items, that is, the set of all available items for selection at step t. Let $\hat{\theta}_{t-1}(X_{t-1})$ be the current provisional ability estimate (using the current response pattern), and $\hat{\theta}_t(X_{t-1}, X_j)$ be the provisional ability estimate obtained when item j is administered at step t (with $j \in S_t$) and response pattern X_{t-1} is updated with item response X_j. Finally, notation j_t^* refers to the item that is selected at the t-th step of the process.

With these notations one can formally introduce the item selection methods. To the best of our knowledge, at least 14 such methods have been introduced to date. The most well known and most popular methods are presented first. Some more recent and specific techniques are then discussed.

The following item selection methods are the most commonly used and found in the CAT literature (e.g., Choi & Swartz, 2009; Magis & Raîche, 2012; van der Linden & Glas, 2010; Wainer, 2000; Weiss, 1983).

1. The *maximum Fisher information* (MFI) criterion. It consists of selecting, among the set of eligible items, the most informative item at the current ability estimate. That is, the selected item j_t^* is such that

$$j_t^* = \arg\max_{j \in S_t} I_j\left(\hat{\theta}_{t-1}(X_{t-1})\right), \qquad (3.1)$$

where $I_j(\theta)$ is the item information function defined by (2.24).

2. The so-called *bOpt* criterion, also sometimes referred to as *Urry's rule* (Urry, 1970). This rule was already introduced in the starting step (Sect. 3.6.1) and is also applicable in the test step as a method for next item selection: the selected item has a difficulty level that is closest to the current ability estimate. That is,

$$j_t^* = \arg\min_{j \in S_t} \left|\hat{\theta}_{t-1}(X_{t-1}) - b_j\right|. \qquad (3.2)$$

Note that this method is not available for polytomous items since there is no one-to-one correspondence between the item difficulty level and the IRT parameters of polytomous models. Moreover, under the Rasch model the item information function (2.24) is maximized whenever $\theta = b_j$, that is, when the ability level is equal to the item difficulty. This implies that under this model, MFI and *bOpt* selection methods are equivalent.

3. The *maximum likelihood weighted information* (MLWI) criterion (Veerkamp & Berger, 1997). Instead of maximizing the (Fisher) information function itself, the MLWI aims at weighting it by the likelihood function of the currently administered response pattern. This was introduced to compensate for the fact that MFI relies only on the ability estimate, which can be severely biased at early stages of a CAT. Weighting the information function appropriately, and integrating it across ability levels, allows reducing the impact of such a bias. More precisely, the selected item maximizes the area under the likelihood-weighted information function:

$$j_t^* = \arg\max_{j \in S_t} \int_{-\infty}^{+\infty} L(\theta|X_{t-1}) \, I_j(\theta) \, d\theta, \qquad (3.3)$$

where $L(\theta|X_{t-1})$ is defined by (2.19) with the notation being expanded to emphasize that the likelihood is computed on the basis of the current response pattern X_{t-1}.

4. The *maximum posterior weighted information* (MPWI) criterion (Linden, 1998a) is similar to the MLWI criterion, except that the information function is weighted by the posterior distribution of ability instead of its likelihood function. That is, for any prespecified prior distribution $f(\theta)$ (for instance standard normal), then

$$j_t^* = \arg\max_{j \in S_t} \int_{-\infty}^{+\infty} f(\theta) \, L(\theta|X_{t-1}) \, I_j(\theta) \, d\theta. \qquad (3.4)$$

5. The *Kullback-Leibler (KL)* divergency criterion (Barrada, Olea, Ponsoda, & Abad, 2009; Chang & Ying, 1996). The KL is a global information measure that evaluates the item discrimination capacity between any possible pairs of trait levels. In this CAT context, KL information for item j at the current ability estimate $\hat{\theta}_{t-1}(X_{t-1})$ and the true ability level, say θ_0, can be written as (Barrada et al., 2009; Chang & Ying, 1996; Magis & Barrada, 2017; van der Linden & Pashley, 2010):

$$KL_j\left(\theta_0 \| \hat{\theta}_{t-1}(X_{t-1})\right) = \sum_{k=0}^{K_j} P_{jk}\left(\hat{\theta}_{t-1}(X_{t-1})\right) \log \left[\frac{P_{jk}\left(\hat{\theta}_{t-1}(X_{t-1})\right)}{P_{jk}(\theta_0)}\right].$$

(3.5)

Because the true ability level θ_0 is unknown, Chang and Ying (1996) proposed weighting the KL information (3.5) by the likelihood function $L(\theta|X_{t-1})$ and to integrate it with respect to θ_0, yielding the KL information criterion

$$KL_j\left(\hat{\theta}_{t-1}(X_{t-1})\right) = \int_{-\infty}^{+\infty} L(\theta|X_{t-1})\, KL_j\left(\theta \| \hat{\theta}_{t-1}(X_{t-1})\right) d\theta.$$

(3.6)

The next item to be selected using KL criterion is the one with maximum KL information, that is,

$$j_t^* = \arg \max_{j \in S_t} KL_j\left(\hat{\theta}_{t-1}(X_{t-1})\right).$$

(3.7)

6. The *posterior Kullback-Leibler* (KLP) criterion (Barrada et al., 2009; Chang & Ying, 1996) follows the same approach as KL criterion, except that in (3.6) the KL information is now weighted by the posterior distribution of ability instead of its likelihood function only:

$$KLP_j\left(\hat{\theta}_{t-1}(X_{t-1})\right) = \int_{-\infty}^{+\infty} f(\theta)\, L(\theta|X_{t-1})\, KL_j\left(\theta \| \hat{\theta}_{t-1}(X_{t-1})\right) d\theta.$$

(3.8)

The next item to be selected with KLP is the one that maximizes (3.8):

$$j_t^* = \arg \max_{j \in S_t} KLP_j\left(\hat{\theta}_{t-1}(X_{t-1})\right).$$

(3.9)

7. The *maximum expected information* (MEI) criterion (van der Linden, 1998a). Instead of maximizing (or minimizing) some objective function that depends on the current response pattern and ability estimate, MEI approach aims at maximizing the expected information that will arise when the next item will be selected and administered. Expectation is computed across the different possible item responses. Thus, the selected item has the following property:

$$j_t^* = \arg \max_{j \in S_t} \sum_{k=0}^{K_j} I_j\left(\hat{\theta}_t(X_{t-1}, k)\right) P_{jk}\left(\hat{\theta}_{t-1}(X_{t-1})\right),$$

(3.10)

where $\hat{\theta}_t(X_{t-1}, k)$ is the provisional ability estimate obtained by updating current response pattern X_{t-1} with response category k for item j, and $P_{jk}\left(\hat{\theta}_{t-1}(X_{t-1})\right)$ is the probability of choosing response category k for item j, computed at the current ability estimate. Note that two types of information functions can be considered with the MEI criterion: the item information function as defined by (2.24), or the *observed information function* $OI_j(\theta)$, which takes the same mathematical form of (2.24) but without the expectation (see e.g., Magis, 2015b, for further details). It is also worth mentioning that van der Linden and Pashley (2000) introduced the maximum expected posterior weighted information (MEPWI) criterion as a mix of both MEI and MPWI approaches. However, Choi and Swartz (2009) demonstrated that the MEPWI is essentially equivalent to MPWI approach.

8. The *minimum expected posterior variance* (MEPV) criterion. In the same spirit as the MEI criterion, MEPV aims at minimizing the posterior variance of the ability estimate. This is done as follows: for each eligible item and each related response category, the EAP estimate of ability is computed by augmenting the current response pattern with the response category of the candidate item, and the related standard error is derived using (2.35). The EPV is then obtained as the weighted sum of all squared standard errors, weighted by the probability of answering the corresponding response category. In other words, the MEPV criterion selects the next item as follows:

$$j_t^* = \arg\min_{j \in S_t} \sum_{k=0}^{K_j} SE_{EAP}^2\left(\hat{\theta}_t(X_{t-1}, k)\right) P_{jk}\left(\hat{\theta}_{t-1}(X_{t-1})\right). \tag{3.11}$$

9. The *random selection* among available items. This can be considered as the benchmark selection method for simulation studies. It is clearly sub-optimal compared to other techniques but it is sometimes implemented in CAT software, such as **catR**. Though this method should never be considered in real CAT assessments, it is worth comparing simulated results to the random selection technique in order to quantify the improvement in ability estimation and precision with a more suitable selection approach.

Some more recent methods emerged and are worth being mentioned too (Magis & Barrada, 2017).

10. The so-called *thOpt* procedure (Barrada, Mazuela, & Olea, 2006; Magis, 2013) aims at computing, for each eligible item, the optimal ability level that maximizes its information function, and then select as the next item the one whose optimum ability level is closest to the current CAT ability estimate. In other words, denote θ_j^* as the optimum ability level for item j, that is,

$$\theta_j^* = \arg\max_\theta I_j(\theta). \tag{3.12}$$

Then, the next item selected is

$$j_t^* = \arg \min_{j \in S_t} \left| \theta_j^* - \hat{\theta}_t(X_{t-1}, k) \right|. \tag{3.13}$$

Formulas for the optimal ability level θ_j^* were derived by Birnbaum (1968) for the 3PL model and by Magis (2013) for the 4PL model. So far there is no equivalent value for polytomous IRT models, so this procedure is only available for dichotomous IRT items.

11. The *progressive* method (Barrada, Olea, Ponsoda, & Abad, 2008, 2010; Revuelta & Ponsoda, 1998) was introduced to decrease item overexposure at the early stages of CAT and is performed as follows. The item selected must maximize an objective function that is the weighted sum of two components: a random component drawn from some accurate uniform distribution, and an information component based on the item information function. More precisely, the chosen item is such that

$$j_t^* = \arg \max_{j \in S_t} \left| (1 - W) R_j + W I_j \left(\hat{\theta}_t(X_{t-1}, k) \right) \right|, \tag{3.14}$$

where W is an accurate weight and R_j is a random number drawn into the interval $\left[0; \max_{j \in S_t} I_j \left(\hat{\theta}_t(X_{t-1}, k) \right) \right]$. The principle behind choosing the weight W is as follows: At the start of the CAT, W is small, so that the optimum function in (3.14) is primarily made of the random component and there is little influence from the information function. As the test goes on, W increases so that the importance of item information increases while the random component decreases. At the end of the CAT, almost always MFI is computed. Optimum methods to select the weight W depend on the stopping rule (see Sect. 3.6.4) and involve an *acceleration parameter*, which controls the rate of increase of W across the CAT. Further details can be found in Barrada et al. (2008), Magis and Barrada (2017) and McClarty, Sperling, and Dodd (2006).

12. The *proportional* method (Barrada et al., 2008, 2010; Segall, 2004) is a two-step stochastic selection approach. First, the probability of each eligible item selected is computed as a function of item information functions for all eligible items: for any eligible item $l \in S_t$,

$$Pr(j_t^* = l) = \frac{I_l \left(\hat{\theta}_t(X_{t-1}) \right)^H}{\sum_{j \in S_t} I_j \left(\hat{\theta}_t(X_{t-1}) \right)^H}. \tag{3.15}$$

The cumulative distribution function of these probabilities (3.15) is then derived, and a random value between 0 and 1 is generated. The next item administered is the one identified by the random number inserted into the cumulative probability distribution. Parameter H is the acceleration parameter and acts similarly to that for the progressive method. Specific formulas for H are available in Barrada et al. (2008, 2010) and Magis and Barrada (2017).

13. Finally, the *global-discrimination index* (GDI) was recently introduced by Kaplan, de la Torre, and Barrada (2015). It consists of computing a CAT form of the global discrimination index commonly used in cognitive diagnosis modeling, yielding the emerging field of *cognitive diagnostic computerized adaptive testing* (CD-CAT; see Hsu, Wang, & Chen, 2013; Wang, 2013; Wang, Chang, & Huebner, 2011). The next item chosen maximizes this GDI. Its full formula, as well as further details, are available in Kaplan et al. (2015).
14. For completeness, the GDI has a Bayesian version, called *posterior global-discrimination index* (GDIP), which allows computing the GDI by introducing the prior distribution of ability (Kaplan et al., 2015).

As one can see, many different item selection rules are available. Most operational CAT programs rely on the simple MFI rule for the sake of simplicity and computational ease, but actually each rule has its own advantages and drawbacks. Comparative studies among some of these methods are available in Barrada et al. (2009), Chang and Ying (1996), Choi and Swartz (2009), Segall (2004) and van der Linden and Glas (2010).

3.6.4 Stopping Step

The stopping step sets the parameters for stopping the adaptive administration of items. Four main stopping rules are commonly considered (van der Linden and Glas, 2010), referred to as (a) the length criterion, (b) the precision criterion, (c) the classification criterion, and (d) the information criterion.

The *length criterion* sets the total number of items to be administered, and the CAT process stops when this number of items have been administered. Longer tests increase the precision of the ability estimates, but shorter tests can also be useful for investigating specific issues at early stages of CAT (Rulison & Loken, 2009). Making use of a fixed CAT length ensures all test takers are receiving exactly the same number of items, but at the cost of various levels of precisions for the ability estimates.

The *precision criterion* stops the CAT when the ability level reaches a predefined level of precision. In other words, the CAT stops when the provisional ability level has corresponding standard error smaller than or equal to the pre-specified threshold. Since longer tests yield greater precision (and hence smaller standard errors), this criterion is most often met when administering sufficiently long tests. It is recommended when one wants to ensure minimum precision to all test takers, but this will probably imply various lengths of CAT tests delivered to the test takers, depending on how informative the item bank is for the estimated ability levels.

The *classification criterion* is used when testing for skill mastery. The main goal is to determine whether the test taker has an ability level greater or less than the ability level indicating skill mastery. In practice, this mastery level is fixed at some accurate ability threshold, and the precision criterion consists of comparing the provisional confidence interval of the current ability estimate to this threshold. If this confidence interval overlaps with the threshold, there is not enough certainty about the final classification (mastering or non-mastering) of the test taker. On the other hand, if the confidence interval does not cover the classification threshold, then the test taker can be classified as mastering the skill (if the threshold is lower than the lower confidence limit) or not mastering the skill (if the threshold is higher than the upper confidence limit). This process can obviously be extended to more than one classification threshold.

Finally, the *information criterion* focuses on the amount of information that is carried by each item at the provisional ability estimate. A necessary condition for the CAT to continue is that the remaining eligible items share enough information to make the total information increase significantly. For this rule, the threshold is the minimum information carried by at least one of the eligible items. If, at the provisional ability estimate, all eligible items have information values smaller than the predefined threshold, the CAT stops. This criterion can be used to avoid administering items that do not carry enough information to the ability estimation and related precision.

Although most of the CAT programs focus on a single stopping rule, it is possible to consider several stopping rules simultaneously and to force the CAT process to stop as soon as at least one of the rules is satisfied. The "multiple stopping rule" approach has advantages, for instance it can ensure the CAT will stop before the total item bank is administered. This option is available in software such as **Firestar** (Choi, 2009) and **catR**.

3.6.5   Final Step

The final step returns the final estimate of the test taker's ability level, using the full response pattern from the adaptive test. Any of the available ability estimators can be considered in this final step. One can use the same estimator from the test step, but one can also combine different estimators in the test and the final steps, e.g., using a Bayesian estimator in the first steps of the process (to avoid the issue of infinite estimates with all-correct or all-incorrect patterns) and, using the traditional ML estimator at the final stage. Reporting additional information from the CAT, for instance the final standard error value and corresponding confidence interval for the ability, is also possible at this final stage.

3.7   Content Balance, Exposure and Security

The practical use of CAT can be a quite complicated process. Since the psychometric, content balance and security objectives for a pool tend to interact with the available item resources, emphasizing one of the objectives in the item selection process generally comes at the expense of the other two. Testlets (i.e., items based on same stimuli) in the pool can also reduce the effectiveness of the item selection methods (Steffen, personal communication, November 10, 2016). The larger the item pool, the better for a CAT administration with content balance and exposure control. But, in some cases, item banks can never be too large. This is because when the frequency of test administration increases, the frequency of each item getting administered increases too. For some operational tests, there are only a limited number of items in the item bank, or there are limited items in certain specified content areas in the item bank due to the many constraints for the categories.

According to test assembly requirements, the item selection must meet certain content coverage requirements to have a balanced content representation for each CAT. This is referred to as *content balancing*, i.e., each CAT must be balanced with respect to some predefined percentages of items coming from each category (Kingsbury & Zara, 1989). Forcing the CAT content to be balanced ensures that at least some percentage of the test items come from each category of items.

There are methodologies to control content balancing including the weighted deviations ATA method (Swanson & Stocking, 1993), the shadow test approach using ATA to balance content (van der Linden & Glas, 2010), as well as the methods by Leung, Chang, and Hau (2003) and Riley, Dennis, and Conrad (2010). They require: (a) an intrinsic classification of items into subgroups of targeted areas; (b) a set of relative proportions of items to be administered from each subgroup. With these elements, Kingsbury and Zara (1989) proposed a simple method, sometimes referred to as the *constrained content balancing* method (Leung et al., 2003):

1. At each step of the CAT process, compute the current empirical relative proportions for each category.
2. Determine the category with largest difference between the theoretical relative proportion and its empirical value.
3. Select the next item from this subgroup and return to step 1.

When some items in an item bank are selected multiple times, it can become problematic if the re-used items get exposed more frequently than they should be. The rate of exposure of the re-used items is called the *item exposure rate*. One reason the items might be over-exposed is that these items are very informative at the average ability level, thus they might be selected more frequently than other items (Davey & Parshall, 1999; van der Linden, 1998b).

If the item exposure rate is too high, then test security becomes a serious concern for the test. If an item gets over-exposed, then pre-knowledge about this item can become available and create unfairness for some test takers. The over-exposed items could also perform differently, give misleading information, and cause issues for the ability estimate. Eliminating those over-exposed items could be costly due to the many processes in developing and calibrating them for the item bank. For these reasons, it is important to ensure that items are not administered too frequently (Chang & Ying, 1996; Stocking & Lewis, 1998).

There are methodologies to control for item exposure. These include selection of more than one optimal item in the neighborhood of the current ability estimate, and a random selection is made among these items for choosing the next administered one: this is the *randomesque method* (Kingsbury & Zara, 1989, 1991). Other commonly used methods include the 5-4-3-2-1 technique (Chang & Ansley, 2003; McBride & Martin, 1983), the Sympson and Hetter method (Hetter & Sympson, 1997), the Davey and Parshall approach (Davey & Parshall, 1999; Parshall, Davey, & Nering, 1998), the Stocking and Lewis conditional and unconditional multinomial methods (Stocking & Lewis, 1998), and the recent method of maximum priority index (Cheng & Chang, 2009). Note finally that some item selection methods were designed to incorporate item exposure control, for instance the proportional and progressive methods (Barrada et al., 2008, 2010) described in Sect. 3.6.3.

3.8   CAT with Regression Trees

Since current CATs rely heavily on IRT, when the samples are small or IRT assumptions are violated, e.g., in a multidimensional test, CATs do not perform well. As an alternative approach, the tree-based CAT algorithm (Yan et al., 2004) was introduced and seems to perform better than or as well as an IRT-based CAT in a multidimensional situation.

A tree-based CAT uses regression tree methods introduced by Friedman, Stone, and Olshen (1984). These are recursive partitioning methods for predicting continuous dependent variables (regression) and categorical variables (classification), and are subsequently referred to as *classification and regression trees* (CART). Given the limitations of current IRT-based applications of CATs including the strong model assumptions and practical calibration sample sizes, Yan et al. (2004) introduced the tree-based approach to CAT. Figure 3.2 is an example of an item-level CAT using a regression tree. The first item administrated is Item 31. If a test taker answers Item 31 correctly, the next item administered will be a more difficult item, Item 27. If a test taker answers Item 31 incorrectly, the next item administered will be an easier item, Item 28. The adaptive test continues in this fashion until the last item is administered. If the last item, e.g., Item 8, is answered correctly, the final estimated score will be 35 (on a scale of 0–60) for this 8-item test.

In a tree-based CAT, item scores refer to the correct responses to the items. The criterion score is the total number correct score for a test consisting of all items. The item scores also determine the splitting of the current sample of test takers into two sub-samples to be administered to the easier and more difficult items at the next stage. Tree-based CAT can be seen as a prediction system that routes test takers efficiently to the appropriate groups, based on their item scores at each stage of testing. It predicts their total scores based on the paths they take and the items they answer, without introducing latent traits or true scores. The detailed tree-based CAT algorithm can be found in Yan, Lewis, and von Davier (2014b).

Yan et al. (2004) showed that when sample sizes are large enough and the adaptive test is long enough, the IRT-based CAT produced consistently better estimates of true scores than the tree-based CAT in a one-dimensional example.

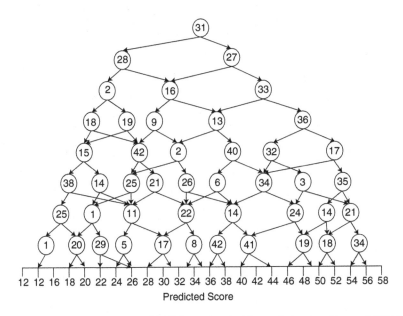

Fig. 3.2 An example of a tree-based CAT *[reproduced with permission from Yan et al. (2014)]*

However, they found that in the early stages of testing, the maximum likelihood estimates from the IRT-based CAT behave very poorly compared to those from the tree-based CAT. In particular, when the items are two-dimensional, the tree-based CAT clearly provided a better prediction than the IRT-based CAT at all test lengths. This suggests that a regression tree performs much better at selecting the first few items on an adaptive test and provides more efficient predictions such as in a "routing test." Therefore, a nonparametric tree-based approach to CAT may be a viable solution when IRT assumptions are not satisfied. Such an approach provides optimal routing and efficiency while scoring.

3.9 Final Comments

CAT is a modern approach for assessment and test administration. It requires the development and calibration of a suitable item bank and an efficient computer interface for test administration, data collection and output production. The latter objective cannot be reached without an efficient software package that encompasses all facets of IRT-based CAT, from ability estimation to next item selection and the computation of the information function. The R package **catR** contains these elements and the purpose of the next chapter is to provide a detailed description of its architecture and functioning.

Chapter 4
Simulations of Computerized Adaptive Tests

In this chapter, we describe the R package **catR** that contains most CAT options and routines currently developed. Its general architecture is presented and most important R functions are explained and illustrated. Focus is put on detailing the input arguments and output values of the main functions, as well as the relationships between them and their accurate use in adaptive and non-adaptive contexts.

4.1 The R Package catR

Computerized adaptive testing requires an accurate software solution for practical purposes. Fundamental CAT research relies primarily on simulated data that allow comparing various CAT scenarios and options with a large set of generated situations. Practical or operational CAT applications, on the other hand, are in need of efficient, stable, fast, attractive, user-friendly and—whenever possible—open source solutions and algorithms to administer adaptive tests on a large scale to potential test takers.

The R package **catR** (Magis & Barrada, 2017; Magis & Raîche, 2012) was developed to allow easy and flexible generation of response patterns under various CAT scenarios. Currently working under dichotomous and polytomous unidimensional IRT models, it covers all facets of CAT, including various item selection rules, ability estimators, and stopping rules. All these aspects are described in detail in the following sections. However, before entering into a complete description of the package and its functionalities, it is essential to set up the item banks in a convenient way, so that all internal R functions can work accurately.

© Springer International Publishing AG 2017
D. Magis et al., *Computerized Adaptive and Multistage Testing with R*, Use R!,
https://doi.org/10.1007/978-3-319-69218-0_4

4.2 Item Bank and Structure

As discussed in Chap. 3, every CAT program requires a suitable item bank to
perform item selection and administration and the latter must be pre-calibrated using
an appropriate IRT model. This pre-calibration step is assumed to be performed
outside of the package's use so only the calibrated item bank must be provided,
together with the IRT model that was chosen for calibration. The current version of
catR supports item banks that were calibrated under the following unidimensional
IRT models:

- the dichotomous 1PL (2.4), 2PL (2.5), 3PL (2.6) and 4PL (2.7) models,
- the GRM (2.10) and MGRM (2.11) polytomous difference models,
- the PCM (2.14), GPCM (2.15), RSM (2.16) and NRM (2.17) polytomous divide-
  by-total models.

Whatever the selected IRT model (dichotomous or polytomous), the item bank
is a matrix with one row per item and as many columns as required to contain the
IRT parameters. For dichotomous IRT models, four columns must be set, each of
them coding for one of the parameters from the 4PL model (2.7) (in this order:
discrimination, difficulty, lower asymptote, upper asymptote). Note that all four
columns must be provided even if a simpler model is considered. For instance, to
set up an item bank under the 2PL model the last two columns must contain zeros
and ones respectively.

In order to further illustrate the R functions of the package, let us create a small
and artificial item bank of 5 items, assumed to having been calibrated under the 2PL
model. Items have discrimination parameters $(0.8, 0.9, 1.0, 1.1, 1.2)$ and difficulty
levels $(0.0, -0.5, 1.1, -0.8, 0.7)$ respectively. The R object that contains the item
bank is called *it.2PL*.

```
R> a <- c(0.8, 0.9, 1.0, 1.1, 1.2)
R> b <- c(0.0, -0.5, 1.1, -0.8, 0.7)
R> it.2PL <- cbind(a, b, c = 0, d = 1)
```

This item bank has the form:

```
R> it.2PL
          a      b  c  d
[1,]  0.8   0.0  0  1
[2,]  0.9  -0.5  0  1
[3,]  1.0   1.1  0  1
[4,]  1.1  -0.8  0  1
[5,]  1.2   0.7  0  1
```

Table 4.1 Maximum number of columns and order of item parameters for all possible polytomous IRT models with $K_j + 1$ (or $K + 1$) response categories

Model	Max. columns	Item parameters
GRM	$1 + \max_j K_j$	$(\alpha_j, \beta_{j1}, \ldots, \beta_{j,K_j})$
MGRM	$2 + K$	$(\alpha_j, b_j, c_1, \ldots, c_K)$
PCM	$\max_j K_j$	$(\delta_{j1}, \ldots, \delta j, K_j)$
GPCM	$1 + \max_j K_j$	$(\alpha_j, \delta_{j1}, \ldots, \delta j, K_j)$
RSM	$1 + K$	$(\delta_j, \lambda_1, \ldots, \lambda_K)$
NRM	$2 \max_j K_j$	$(\alpha_{j1}, c_{j1}, \ldots, \alpha_{j,K_j}, c_{j,K_j})$

With polytomous items, however, the number of columns depends on both the model itself and the number of response categories. As was discussed in Sect. 2.2.2, some models such as GRM, PCM, GPCM and RSM allow for different numbers (K_i) of item response categories, while others such as MGRM and RSM assume the same number (K) of responses for all items. To overcome this problem, the number of columns must be equal to the maximal number of item parameters, and for items with fewer parameters than the maximum, the corresponding row in the bank matrix must be filled out with *NA* values (for "not available"). Table 4.1 displays, for each polytomous IRT model, the maximum number of columns and the appropriate ordering of the item parameters.

One small item bank will be created in order to illustrate one of these polytomous structures. It is called it.GRM and is a small set of five items assumed to be calibrated under the GRM. The first and second items have four response categories, the third and fourth ones have five categories each, and the fifth one has only three categories. The (artificial) parameters are introduced into R as follows:

```
R> it.GRM <-rbind(c(1.169, -0.006, 0.764,
  2.405, NA),
+              c(1.052, -1.148, -0.799, -0.289, NA),
+              c(0.828, -0.892, -0.412, -0.299, 0.252),
+              c(0.892, -1.238, -0.224, 0.377, 0.436),
+              c(0.965, 0.133, 0.804, NA, NA))
```

and the resulting item bank is:

```
R> it.GRM
        [,1]    [,2]    [,3]    [,4]   [,5]
[1,] 1.169 -0.006  0.764  2.405     NA
[2,] 1.052 -1.148 -0.799 -0.289     NA
[3,] 0.828 -0.892 -0.412 -0.299 0.252
```

(continued)

```
[4,]  0.892  -1.238  -0.224   0.377  0.436
[5,]  0.965   0.133   0.804     NA     NA
```

4.3   General Architecture of catR

The package **catR** is built with a hierarchical structure, starting from low-level, basic functions for IRT modeling and scoring, to top-level functions that embed all CAT aspects. This structure is schematically depicted in Fig. 4.1 and detailed hereafter.

First, *basic IRT functions* perform basic IRT computations such as item response probabilities, item information functions, and so on. They serve as fundamental functions for the second layer of functions, referred to as the *IRT level*, useful to generate item parameters and item responses and to perform IRT scoring (i.e., ability estimation and standard error computation). The third layer is the *CAT level* and encompasses all methods for intermediate technical computations of CAT related quantities (such as expected information, expected posterior variance, Kullback-Leibler information...), the selection of the first item(s) (to start the CAT) and of the next item (during the CAT). Eventually, *top-level functions* perform full CAT generation of either one or several response patterns under pre-specified CAT scenarios. Each group of functions is described hereafter with details about input and output arguments.

Note that **catR** also holds several routine functions to e.g., check the format of some input lists and return warning messages if some of their arguments are

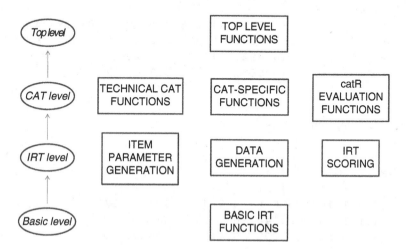

Fig. 4.1 General structure of the **catR** package

incorrectly specified. These functions, merged into the group of *catR evaluation functions*, add an extra layer of stability and provide insightful information to the test user in case of misspecified R code. Despite their intrinsic interest for the test administrator, such test functions will not be further described in this chapter.

4.4 Basic IRT Functions

The first basic IRT function of interest is called Pi() and computes the item category probabilities and its first, second and third derivatives. It takes as input argument the underlying ability level th, the matrix of item parameters it and optionally the acronym of the polytomous IRT model as argument model. By default, model is set to NULL so that dichotomous IRT model is assumed. In addition, the scaling constant D can be set to another value than the default 1 with the argument D. As an example, with an ability level of zero and the 2PL item bank it.2PL defined in Sect. 4.2,

```
R> Pi(th = 0, it = it.2PL)
```

returns the following list of probabilities ($Pi), first ($dPi), second ($d2Pi) and third ($d3Pi) derivatives:

```
$Pi
[1] 0.5000000 0.6106392 0.2497399 0.7068222
 0.3015348

$dPi
[1] 0.2000000 0.2139831 0.1873699 0.2279470
 0.2527339

$d2Pi
[1]  0.00000000 -0.04261486  0.09378241 -0.10371792
[5]  0.12038132

$d3Pi
[1] -0.06400000 -0.07393298 -0.02327495 -0.06711909
[5] -0.09595897
```

The second basic function is Ii() and computes the item information function as well as its first and second derivatives. It takes the same arguments as function Pi() and returns a similar list with arguments Ii, dIi and d2$Ii. As an illustration with the 2PL item bank,

```
R> Ii(th = 0, it = it.2PL)
```

returns

```
$Ii
[1] 0.1600000 0.1925848 0.1873699 0.2507417
 0.3032806

$dIi
[1]  0.00000000 -0.03835337  0.09378241 -0.11408971
[5]  0.14445758

$d2Ii
[1] -0.05120000 -0.06653968 -0.02327495 -0.07383099
[5] -0.11515077
```

It is also possible to compute the observed information function using the OIi() function. In addition to the arguments th and it (and optionally model for polytomous IRT models), it takes as input the vector x of item responses for the whole set of items defined in the matrix it of item parameters. For instance, the vector $(0, 1, 0, 0, 1)$ can be used to compute observed information values with the 2PL matrix as follows:

```
R> x <- c(0, 1, 0, 0, 1)
R> OIi(th = 0, it = it.2PL, x = x)
```

and the output is a vector of observed information values:

```
[1] 0.1600000 0.1925848 0.1873699 0.2507417
[5] 0.3032806
```

Note that the previous output is identical to the one from the function Ii() above. This is because under the 2PL model, observed and expected information functions are totally equivalent (see e.g., Bradlow, 1996).

The function Ji() returns the value of the $Ji(\theta)$ function that stands in the definition (2.26) of the weighted likelihood estimator (Warm, 1989). It takes the same arguments as function Ii() and returns a list with values of the $Ji(\theta)$ function ($Ji) and its first derivative ($dJi).

4.5   IRT-Level Functions

This sections briefly outlines the three groups of R functions that form the second layer of **catR** functions: the *item parameter generation* functions, the *data generation* functions and the *IRT scoring* functions.

4.5.1   Item Parameter Generation

The package **catR** has three functions allowing for generation of one's own sets of item parameters and response matrices. This is particularly useful to simulate specific item banks under predefined IRT models, and to run CAT simulations by generating random item responses for given item parameters and ability levels. This section focuses on item parameter generation, while next section is devoted to data (i.e., item response) generation.

The function genDichoMatrix() creates a matrix of item parameters for dichotomous IRT models. The input arguments are: items to specify the number of items in the generated matrix; model to define which dichotomous IRT model will be considered (either "1PL", "2PL", "3PL" or "4PL"); seed to fix the random seed for parameter value generation; and the arguments aPrior, bPrior, cPrior and dPrior to set the parent distributions of related parameters. The latter arguments are set by vectors of three components, the first one holding the acronym of the distribution, and the last two components defining the parameters of the distribution. Possible combinations of these parent distributions are:

1. the *normal distribution* set by the vector c("norm", mu, sigma) where mu and sigma specify the mean and standard deviation, respectively;
2. the *log-normal distribution* set by the vector c("lnorm", mu, sigma) where mu and sigma specify the mean and standard deviation of the log-normal distribution, respectively;
3. the *uniform distribution* set by the vector c("unif", a, b) where a and b specify the lower and upper limits of the uniform interval, respectively;

4. the *Beta distribution* set by the vector c("beta", alpha, beta) where alpha and beta specify the corresponding parameters of the $B(\alpha, \beta)$ distribution.

The aPrior argument can be defined using the normal (by default), the log-normal and the uniform distributions; the bPrior argument with normal (default) and uniform distributions; and the cPrior and dprior arguments with either the uniform (default) or Beta distributions.

To illustrate this function, let us consider the creation of a set of five items under the 3PL model with the following parent distribution: the log-normal distribution with parameters 0 and 0.12 for discrimination levels; the standard normal distribution for difficulty levels, and the $B(5, 17)$ distribution for lower asymptotes. The corresponding code and output are given below:

```
R> genDichoMatrix(items = 5, model = "3PL",
+                 aPrior = c("lnorm", 0, 0.12),
+                 bPrior = c("norm", 0, 1),
+                 cPrior = c("beta", 5, 17),
+                 seed = 1)

          a          b          c d
1 0.9062354 -0.6264538 0.43894249 1
2 1.0602360  0.1836433 0.27022042 1
3 1.0926424 -0.8356286 0.16873677 1
4 1.0715367  1.5952808 0.05708731 1
5 0.9640168  0.3295078 0.37145663 1
```

The genDichoMatrix() holds a final argument called cbControl that allows specifying an underlying structure to the item bank for content balancing purposes. In practice, cbControl must be provided a list with two arguments: $names contains the different names of the subgroups of items, and $props contains a vector of numeric values that specify the proportions of items to be allocated to each subgroup. For instance, setting $names to c("a", "b", "c") and $props to c(0.2, 0.4, 0.4) will allocate approximately 20% of the items to the subgroup "a", 40% of the items to the subgroup "b" and 40% of the items to the subgroup "c". This situation is embedded in the previous illustrative example as follows:

```
R> genDichoMatrix(items = 5, model = "3PL",
+                 aPrior = c("lnorm", 0, 0.12),
```

(continued)

```
+                        bPrior = c("norm", 0, 1),
+                        cPrior = c("beta", 5, 17),
+                        seed = 1,
+                        cbControl = list(
+                        names = c("a", "b", "c"),
+                        props = c(0.2, 0.4, 0.4)))

           a           b           c d Group
1 0.9062354 -0.6264538 0.43894249 1     b
2 1.0602360  0.1836433 0.27022042 1     c
3 1.0926424 -0.8356286 0.16873677 1     a
4 1.0715367  1.5952808 0.05708731 1     b
5 0.9640168  0.3295078 0.37145663 1     c
```

The output matrix contains an additional column with the membership of each item
to each subgroup.

Polytomous item banks can be generated using a similar approach with the
genPolyMatrix() function. Arguments items, seed and cbControl are
identical to the ones of genDichoMatrix(), while the model argument
now takes one of the acronyms of the available polytomous IRT models listed
in Table 4.1. In addition, arguments nrCat specifies the maximum number of
response categories per item (these numbers are randomly drawn between 2 and
nrCat), and argument same.nrcat permits forcing all items to have the same
number of response categories. In case of modified graded response and rating scale
models, the latter argument is ignored since these polytomous models impose the
same number of response categories across items.

Currently the parent distributions of polytomous IRT models are fixed as
follows:

1. the log-normal distribution with parameters 0 and 0.1225 for the α_j parameters
   of the GRM, MGRM and GPCM and for the α_{jk} parameters of the NRM;
2. the standard normal distribution for all other parameters.

As an illustration, the following code generates a set of six items under the
GRM with at most four response categories, and subdivided into three groups of
approximately equal size:

```
R> genPolyMatrix(items = 6, model = "GRM",
+               nrCat = 4, seed = 1,
+               cbControl = list(
```

(continued)

```
+                         names = c("a", "b", "c"),
+                         props = c(1/3, 1/3, 1/3)))

   alphaj betaj1 betaj2 betaj3 Group
1   1.216 -0.305  1.512     NA     a
2   1.041 -0.621  0.390     NA     a
3   0.904 -2.215 -0.045  1.125     b
4   1.062 -0.016  0.821  0.944     c
5   1.095  0.594  0.919     NA     c
6   1.073 -1.989  0.075  0.782     b
```

4.5.2   Data Generation

Once item parameters (either generated from functions described in Sect. 4.5.1 or from real calibration) and true ability levels are available, item responses can be generated with the `genPattern()` function. Input arguments are: a vector of ability levels `th`, a matrix of item parameters `it`, the optional setting of the polytomous IRT model (with argument `model`), the value of the scaling constant D (`D`), and the value to fix the random seed (`seed`). The output is a matrix with as many rows as the number of ability levels specified in the `th` argument, and as many columns as the number of items in the `it` argument. As an illustration, let us generate a matrix of responses for 10 ability levels evenly spaced from -2 to 2 and the matrix `it.GRM` created in Sect. 4.2. This response matrix will be stored in the object `data` for further use.

```
R> data <- genPattern(th = seq(-2, 2, length = 10),
+                     it = it.GRM, model = "GRM",
+                     seed = 1)
R> data

      [,1] [,2] [,3] [,4] [,5]
[1,]     0    0    0    2    1
[2,]     0    0    0    0    0
[3,]     0    0    0    0    2
[4,]     1    3    4    0    0
[5,]     0    3    0    1    0
[6,]     0    3    0    2    2
```

(continued)

[7,]	0	3	4	4	0
[8,]	2	3	4	4	0
[9,]	0	2	4	4	2
[10,]	3	3	4	3	2

4.5.3 IRT Scoring

In CAT designs using an IRT framework, ability estimation is mandatory both during and at the end of the test. Moreover, some item selection procedures (such as maximum expected information or minimum expected posterior variance) require provisional estimation of ability and related standard errors. For those purposes, functions thetaEst() and semTheta() were created to estimate the ability levels and compute their associated standard errors, respectively.

The function thetaEst() takes, as mandatory input arguments, it that provides the matrix of item parameters, and x that holds the item responses. The choice of the scaling constant D is set through argument D and the appropriate polytomous model is referred to by the argument model. The choice of the ability estimator is driven by the argument method, which takes the default value "BM" for Bayes modal estimation. Other possible values are "ML" for maximum likelihood, "WL" for weighted likelihood and "EAP" for expected a posteriori estimation. For the Bayesian estimators (BM and EAP), it is possible to specify the prior distribution with the arguments priorDist and priorPar. Argument priorDist specifies the type of prior distribution and can be either "norm" for the normal distribution (by default), "unif" for the uniform distribution, or "Jeffreys" for Jeffreys' non-informative prior (Jeffreys, 1939, 1946). Argument priorPar, on the other hand, is a vector of two components with the parameters of the prior distribution. For the normal prior, priorPar contains the mean and standard deviation of that distribution, while for the uniform prior it holds the lower and upper values of the range of the uniform density. This priorPar argument is not used if Jeffreys' prior was chosen. Note also that the number of quadrature points for EAP estimation can be defined by the argument parInt. This argument takes three values that define a regular sequence of points: the lower value, the upper value and the number of quadrature points. By default the argument's values are -4, 4 and 33, allowing 33 quadrature points evenly spaced from -4 to 4.

A technical problem can occur in case of constant patterns, i.e., responses that belong all to the first or to the last response category (in case of polytomous IRT models) or simply when all responses are either correct or incorrect (in case of dichotomous IRT models). In this situation, ML estimates become infinite. To overcome that problem, the argument range fixes the maximum range of ability estimation, and is by default equal to $[-4, 4]$. Infinite positive (negative) estimates are then shrunk to equal the upper (lower) bound of that range.

Another way to overcome this issue of constant patterns is to make use of ad-hoc estimates of ability. This is particularly useful at early stages of CAT applications in which only a few item responses are available. Such ad-hoc adjustments consist mostly of stepsize adjustments (Dodd, De Ayala, & Koch, 1995). Two approaches are implemented in **catR** (currently only for dichotomous IRT models): fixed stepsize adjustment in which the ability estimate gets increased or decreased from some fixed step value (to be specified by the user); or variable stepsize adjustment, in which the current ability level is increased or decreased by half the distance between this current value and the largest (or smallest) difficulty level in the item bank.

The `thetaEst()` function allows these stepsize adjustments as follows. First, the argument `constantPatt` must be set to either `"fixed4"`, `"fixed7"` or `"var"` (by default it takes the NULL value so that no stepsize adjustment is performed). Both `"fixed4"` and `"fixed7"` lead to fixed stepsize adjustments with steps of .4 and .7, respectively. The `"var"` value yields variable stepsize adjustment and the range of difficulty levels must then be provided through the argument `bRange`. Finally, the current ability level must also be provided through the `current.th` argument.

Some examples of ability estimation in **catR** are provided below. Using the formerly defined set of items under the GRM, one response pattern is being generated with the `genPattern()` function and true ability level of zero. Then, ability is estimated using, respectively: the BM method (default), the ML method, the EAP method, the WL method, and finally the EAP method with Jeffreys' prior.

```
R> genPattern(th = 0, it = it.GRM, model = "GRM",
+            seed = 1)
R> thetaEst(it = it.GRM, x = x, model = "GRM")
[1] -0.1229094
R> thetaEst(it = it.GRM, x = x, model = "GRM",
+           method = "ML")
[1] -0.2247436
R> thetaEst(it = it.GRM, x = x, model = "GRM",
+           method = "EAP")
[1] -0.1294389
R> thetaEst(it = it.GRM, x = x, model = "GRM",
+           method = "WL")
[1] -0.1972096
R> thetaEst(it = it.GRM, x = x, model = "GRM",
+           method = "EAP",
+           priorDist = "Jeffreys")
[1] -0.2183364
```

Once an ability estimate is obtained, the related standard error (SE) can be computed using the semTheta() function. Several arguments are similar to those of the thetaEst() function: arguments it, x, model, method, priorDist, priorPar, parInt and constantPatt take exactly the same values as previously described. In addition, the ability estimate must be provided through the argument thEst.

It is interesting to note, first, that the response pattern x must be provided only with EAP estimates, as the computation of the standard error does not require this pattern for ML, BM or WL methods. Second, in case of stepsize adjustment of the ability estimate (i.e., when constantPatt is different from the NULL value), the SE value is returned as the infinite value Inf. This is because standard errors cannot be computed with such adjustments, so fixing the returned value to Inf avoids early stops in the CAT process when the stopping rule is the precision criterion.

To illustrate this function, using the previously generated response pattern, the SE values of the estimated abilities are now obtained as follows.

```
R> th.BM <- thetaEst(it = it.GRM, x = x,
+                         model = "GRM")
R> semTheta(thEst = th.BM, it = it.GRM,
+            model = "GRM")
[1] 0.6493841
R> th.ML <- thetaEst(it = it.GRM, x = x,
+                         model = "GRM", method = "ML")
R> semTheta(thEst = th.ML, it = it.GRM,
+            model = "GRM", method = "ML")
[1] 0.8565175
R> th.EAP <- thetaEst(it = it.GRM, x = x,
+                         model = "GRM", method = "EAP")
R> semTheta(thEst = th.EAP, it = it.GRM, x = x,
+            model = "GRM", method = "EAP")
[1] 0.67968
R> th.WL <- thetaEst(it = it.GRM, x = x,
+                         model = "GRM", method = "WL")
R> semTheta(thEst = th.WL, it = it.GRM,
+            model = "GRM", method = "WL")
[1] 0.8556875
R> th.J <- thetaEst(it = it.GRM, x = x,
+                         model = "GRM", method = "EAP",
+                         priorDist = "Jeffreys")
R> semTheta(thEst = th.J, it = it.GRM, x = x,
+            model = "GRM", method = "EAP",
+            priorDist = "Jeffreys")
[1] 0.8928566
```

It is important to notice that missing values in the response pattern can be handled in the ability estimation process (and associated SE computation). Such missing values must be coded as NA in the response pattern and related items are discarded from the estimation process. Furthermore, in the presence of missing data, the response pattern must be provided to the semTheta() function, even if an estimator other than EAP was used.

As an illustration, let us remove the second item response from previously generated vector x (which was a zero value) and replace it by a missing value, and let us see the impact on the BM estimator (other methods can be replicated as well).

```
R> x2 <- x
R> x2[2] <- NA
R> rbind(x,x2)
    [,1] [,2] [,3] [,4] [,5]
x     2    0    1    0    2
x2    2   NA    1    0    2
R> th.BM <- thetaEst(it = it.GRM, x = x,
+                    model = "GRM")
R> th.BM2 <- thetaEst(it = it.GRM, x = x2,
+                     model = "GRM")
R> c(th.BM, th.BM2)
[1] -0.1229094  0.2637767
R> semTheta(thEst = th.BM, it = it.GRM,
+           model = "GRM", x = x)
[1] 0.6493841
R> semTheta(thEst = th.BM2, it = it.GRM,
+           model = "GRM", x = x2)
[1] 0.6907471
```

As one would expect, removing a zero response yields an increase in estimated ability, but also an increase in SE value, since less information is available (one item being discarded from the computations).

4.6 CAT-Level Functions

In this section, two blocks of functions are described: the *technical CAT functions* and the *CAT specific* functions.

Table 4.2 List of technical CAT functions of **catR** for next item selection

Function	Description
EPV()	Computes the expected posterior variance
GDI()	Computes the global-discrimination index (GDI) and posterior global-discrimination index (GDIP)
KL()	Computes the Kullback-Leibler (KL) and posterior Kullback-Leibler (KLP) values
MEI()	Computes the maximum expected information
MWI()	Computes the maximum likelihood weighted information (MLWI) and maximum posterior weighted information (MPWI)

4.6.1 Technical CAT Functions

Technical CAT functions are primarily used to select the next item and perform intermediate calculations at the item level. Such functions are used as internal parts of other, more interesting CAT functions (to be further described below), and hence they are not discussed in detail in this book. Table 4.2 lists all these technical CAT functions together with descriptions. Further information can be found in the corresponding help files from the **catR** package.

4.6.2 CAT-Specific Functions

The three main aspects of adaptive testing are: (a) the selection of the first item(s) to administer; (b) the iterative selection of the next item to administer; and (c) the setting of the stopping rule(s). The latter step is checked internally from top-level functions, while the former two are handled by separate functions in **catR**, namely startItems() and nextItem().

4.6.2.1 The startItems() Function

The selection of the first item(s) is performed by the function startItems(). The basic input is the item bank itself (set by argument itemBank), that is, a matrix with parameters of the available items (in the same format as described in Sect. 4.2), and the type of polytomous IRT model (set by argument model) if necessary. The selection of the first items can be set in three different ways: by letting the user select the first items to administer; by picking them randomly from the bank; or by selecting them optimally according to some pre-specified options (see later).

The forced administration of items is implemented by providing some integer values to the argument fixItems. These values correspond to the numbering (i.e., position) of the selected items in the bank (i.e., number one for the first item, number two for the second item etc.). This argument can reduce to a single value, in this case only one starting item is selected by the user. Once `fixItems` takes a value different from its default `NULL` value, this selection process will be applied (independently of the other arguments).

If the starting items are not selected by the user, they can be drawn randomly from the item bank. In this case, several arguments must be supplied: `seed` fixes (optionally) the random seed for first item selection and `nrItems` specifies how many items must be randomly chosen (by default only one item is selected). The `seed` argument takes the `NULL` default value (so that random selection is disabled) and can take either any real value (to fix the random seed) or `NA` (to allow random selection without fixing the random seed).

If neither forced nor random selection is performed (i.e. if both `fixItems` and `seed` arguments take `NULL` value), the selected items will be chosen as "optimal" for a given set of starting ability levels. These ability levels are set with the `theta` argument as a vector of numeric values, one per item to select. For instance, defining `theta` as the vector `c(-1, 0, 1)` means that three starting items will be chosen, one being "optimal" at ability level -1 and the other two being "optimal" at levels 0 and -1 respectively. Furthermore, optimal selection of the first items is driven by one of the following processes:

1. `"MFI"` (Maximum Fisher information): the most informative item is chosen for each starting ability level.
2. `"bOpt"`: item whose difficulty level is as close as possible to the initial ability level is selected.
3. `"thOpt"`: the selected item is the one for which the ability level that maximizes its Fisher information is as close as possible to the target initial ability level. Such optimal ability levels are available for dichotomous IRT models (Magis, 2013) so this method is currently restricted to dichotomous IRT models.
4. `"progressive"` and `"proportional"` are also available options but they refer directly to the corresponding methods for next item selection and are therefore not further discussed.

Note that currently only the `"MFI"` method is available for polytomous IRT models.

Whatever the optimal criterion, `startItems()` will select, by default, the most optimal item for each pre-specified ability level in the `theta` argument. However, to allow for more randomness in the initial choice of these items and avoid the overexposure of the same initial items across repeated simulations, the *randomesque method* (Kingsbury & Zara, 1989) has been made available at this stage. It consists of identifying first, for each starting ability level, a small set of most optimal items (using the chosen selection rule) and then selecting randomly one of these identified items as the starting one to administer. In this way, one preserves (almost) optimality in starting item selection while introducing some randomness in the process. In practice, the randomesque approach is set up by assigning to the `randomesque` argument, the number of items to optimally select in the first step (by default this argument equals one, so no randomesque selection is performed).

The random selection among these "randomesque" items can be fixed by specifying some numeric value to the random.seed argument.

Finally, it is possible to restrict the item selection to some subset of the item bank. For instance, one could discard some informative items from being selected at the start of a CAT or limit item exposure from previous runs of the simulation process. This can be set with the nAvailable argument. It must be provided a vector of zeros and ones only, with as many components as the number of items in the bank (i.e. one component per row of the itemBank matrix). Each zero value identifies the item as being not available for selection, so that only items flagged with a one in the nAvailable vector can be considered for item selection.

The output of the startItems() function is a list with several arguments, the most important being $items (which returns the selected items) and $par which provides the IRT parameters of the selected items. Other output values are returned but they are basically used to transfer input information in the meta-functions to be described in Sects. 4.7 and 4.8.

Two illustrative examples are now displayed, one with the artificial polytomous it.GRM item bank and one with the artificial dichotomous it.2PL item bank, both generated in Sect. 4.2. First, the two most informative items are selected from the it.GRM matrix with respect to the starting abilities −1 and 1.

```
R> startItems(itemBank = it.GRM, model = "GRM",
+               theta = c(-1, 1))
$items
[1] 2 1

$par
       [,1]    [,2]    [,3]    [,4] [,5]
[1,] 1.052 -1.148 -0.799 -0.289   NA
[2,] 1.169 -0.006  0.764  2.405   NA

$thStart
[1] -1  1

$startSelect
[1] "MFI"
```

It is straightforward to check that items 2 and 1 are the most informative items for ability levels −1 and 1 respectively, using the Ii() function described in Sect. 4.4. For instance, the following code and output highlights that item 2 yields the most information at ability level −1.

```
R> Ii(th = -1, it = it.GRM, model = "GRM")$Ii
[1] 0.2529312 0.3194691 0.1914388 0.2356644
 0.1786138
```

Now, the "bOpt" criterion will be used with the it.2PL matrix, for initial
ability level zero. Given that the first item has a difficulty level equal to zero, we
will discard this item from the selection process. Moreover, randomesque selection
will be set with a randomesque value of two. That is, the two most optimal items
are items 2 and 5 (since their difficulty levels are the closest to zero and item 1 is
discarded from the selection process) and one of these two will be selected randomly
(random seed being fixed to one). The corresponding code and output are:

```
R> startItems(itemBank = it.2PL, theta = 0,
+             startSelect = "bOpt",
+             randomesque = 2, random.seed = 1,
+             nAvailable = c(0, rep(1, 4)))
$items
[1] 2

$par
   a    b    c    d
 0.9 -0.5  0.0  1.0

$thStart
[1] 0

$startSelect
[1] "bOpt"
```

In this example, item 2 was chosen.

4.6.2.2 The nextItem() Function

The second main CAT-specific function is the nextItem() one. It permits
selection of the next item to administer, for a given current ability estimate, a
response pattern and a set of previously administered items. The choice of the
method for next item selection is also a mandatory input argument.

The full item bank must be provided through the argument itemBank and the
set of previously administered items through the argument out. The latter must be

a vector of integer values that refer to the item positions in the bank (or NULL if no item was administered yet); these items are then discarded from the selection process. As usual, the model argument defines the appropriate polytomous IRT model that was used to calibrate the bank (if necessary). The current ability level is specified by the argument theta and the response pattern is supplied by the argument x. The latter must be ordered in the same way as the set of previously administered items that is provided by out. Note that some methods (listed below) do not require the specification of the current ability estimate, the current response pattern, or both. However, the set of previously administered items must always be supplied.

The method for next item selection is set by the argument criterion. Currently 14 methods are available for dichotomous IRT models and some of them can also be used for polytomous IRT models. All these methods are described in Sect. 3.6.3. They are listed below, together with the acronym to be used in the nextItem() function.

1. "MFI": the maximum Fisher information (MFI) criterion,
2. "bOpt": the so-called *bOpt* procedure (*not for polytomous items*),
3. "thOpt": the so-called *thOpt* procedure (*not for polytomous items*),
4. "MLWI": the maximum likelihood weighted information (MLWI),
5. "MPWI": the maximum posterior weighted information (MPWI),
6. "MEI": the maximum expected information (MEI) criterion,
7. "MEPV": the minimum expected posterior variance (MEPV),
8. "KL": the Kullback-Leibler (KL) divergency criterion,
9. "KLP": the posterior Kullback-Leibler (KLP) criterion,
10. "progressive": the *progressive* method,
11. "proportional": the *proportional* method,
12. "GDI": the global-discrimination index (GDI),
13. "GDIP": the posterior global-discrimination index (GDI),
14. "random": the random selection among available items.

Some of these methods require the specification of additional options. First, the ability estimator, its prior distribution and prior parameters (in case of Bayesian estimation), the scaling constant D, the range of ability estimation and the parameters for numerical integration can be provided through the arguments method, priorDist, priorPar, D, range, and parInt, respectively. Further details can be found in Sect. 4.5.3. For the "MEI" method, it is possible to specify which type of information function (observed or expected) will be used, through the argument infoType. For the "progressive" and "proportional" methods, the acceleration parameter can be tuned by the argument AP.

Similarly to the selection of the first items, item exposure can be controlled by the *randomesque* method using the so-called argument randomesque. More precisely, this argument determines how many optimal items must be selected by the chosen method, and the finally administered item is selected randomly among those optimal items.

Eventually, content balancing can also be controlled at this stage (Kingsbury & Zara, 1989). This is performed by the arguments cbControl and cbGroup (which are set to NULL by default, so that no content balancing control is performed). The cbGroup argument is a vector with as many components as the number of items in the bank, and each component holds the name or label of the sub-group the item belongs to. The cbControl argument, on the other hand, is a list with elements $names and $props and hold, respectively, the names and the relative proportions of items of each subgroup (further details and examples can be found in Sect. 4.4 on matrix generation with content balancing structures).

The selection of the next item under content balancing control is done in two steps. First, the optimal sub-group of items is targeted as being either one of the sub-groups from which none of the items was administered yet (in case of several such sub-groups, one of them is chosen randomly), or the sub-group for which the gap between the observed relative proportion of administered items (computed using the vector out of administered items) and the expected relative proportion (set by argument cbControl$props) is maximal. In a second step, the most optimal item among the selected subgroup is chosen for next administration.

The output of the nextItem() function is a list with several elements. Among other elements, $item mentions which item was selected and $par returns the parameters of the selected item. Other elements are returned for proper use in the upcoming functions for CAT simulations, and elements $prior.prop, $post.prop and $cb.prop are returned in case of content balancing control (see below).

Given the important number of possible configurations for next item selection, only one simple illustrative example will be described (other examples can be found in the help file of nextItem() function and in Chap. 5). The 2PL matrix it.2PL will be used. It is assumed that items 1 and 3 were already administered and a correct response and an incorrect responses were recorded, respectively. The MLWI method is chosen to select the next item, which requires only the specification of the itemBank, out and x arguments. Moreover, it is assumed that the first two items belong to the sub-group "a" and the last three items to the sub-group "b". Content balancing control is imposed by requesting theoretical proportions of one third of items from sub-group "a" and two thirds of items from sub-group "b" (which will actually force the selection of the next item in sub-group "b" in this example). The code and output are:

```
R> x <- c(1, 0)
R> out <- c(1, 3)
R> group <- c("a", "a", "b", "b", "b")
R> control <- list(names = c("a", "b"),
+                  props = c(1/3, 2/3))
R> nextItem(itemBank = it.2PL, out = out,
+           x = x, criterion = "MLWI",
```

(continued)

```
+                    cbGroup = group,
+                    cbControl = control)
$item
[1] 5

$par
  a    b    c    d
1.2  0.7  0.0  1.0

$info
[1] 0.3505903

$criterion
[1] "MLWI"

$randomesque
[1] 1

$prior.prop
[1] 0.5 0.5

$post.prop
[1] 0.3333333 0.6666667

$cb.prop
[1] 0.3333333 0.6666667
```

As can be seen from the output, item 5 is eventually selected and the corresponding likelihood-weighted information equals 0.351 (argument $info). Before item selection, the relative proportions of items chosen in groups "a" and "b" (returned in element $prior.prop) were equal to 0.5 (one item selected in each sub-group) and after item selection, these proportions (returned in the $post.prop element) evolve to one third and two thirds (since item 5 is chosen and belongs to sub-group "b"). The expected proportions are returned in the $cb.prop element and are (in this example) identical to the posterior proportions.

4.7   Top-Level Function: randomCAT()

To end up this overview of the **catR** package, the main two top-level functions, namely randomCAT() and simulateRespondents(), are described. Both functions make calls to previously described functions and share all aspects of a

CAT design (selection of the first item(s), next item selection, provisional ability estimation, stopping rule, final ability estimation and summary indexes, item exposure and content balancing control). The main difference between the two functions is that `randomCAT()` allows for generation of a single CAT response pattern, while `simulateRespondents()` results in the multiple generation of many CAT response patterns (with global control for content balancing and item exposure too).

Both functions will be described in terms of their input arguments. This section focuses on `randomCAT()` function and Sect. 4.8 describes the function `simulateRespondents()`. Some examples and illustrations are displayed in the next chapter to illustrate the usefulness of **catR** in CAT simulations.

4.7.1 Input Information

Let us start by listing all input arguments of the `randomCAT()` function in Table 4.3.

The item bank must be supplied through the `itemBank` argument, and the underlying IRT model through the `model` argument (either the acronym of the polytomous IRT model or `NULL` for the dichotomous IRT models). The item bank structure must be identical to those presented in Sect. 4.2.

Two types of CAT simulation can be run: *full simulations* or *post-hoc simulations*. Full simulations imply that the responses to the selected items are generated from the item parameters and the true ability level that is specified by the

Table 4.3 Input arguments for `randomCAT()` function

Name	Function	Type
trueTheta	Sets the true ability level	Numeric
itemBank	Sets the matrix of item parameters	Matrix
model	Sets the type of polytomous IRT model	Model acronym or NULL
responses	Provides item responses for post-hoc simulations	Vector or NULL
genSeed	Fixes the general random seed	Numeric or NULL
cbControl	Sets the options for content balancing control	Appropriate list or NULL
nAvailable	Optionally sets the available items in the bank	Binary vector or NULL
start	Sets the options to select the first item(s)	Appropriate list
test	Sets the options for provisional ability estimation and next item selection	Appropriate list
stop	Sets the options of the stopping rule(s)	Appropriate list
final	Sets the options for final ability estimation	Appropriate list
allTheta	Sets whether all ability estimates must be returned	Logical
save.output	Sets whether output should be saved	Logical
output	Sets the output location, file name and extension	Appropriate vector

trueTheta argument. The genPattern() function is used for that purpose. Post-hoc simulations consist of drawing item responses from a predefined set of responses that was previously collected (during e.g., linear testing) and is provided by the responses argument. In the latter case, responses must be a vector of the same length of the number of items in the bank and the trueTheta argument may not be provided (as it is not used to generate the item responses). If the responses argument is specified as NULL, however, the full CAT simulation is assumed and a true ability level must be provided.

In case of a full CAT simulation, the random drawing of item responses can be fixed through the genSeed argument by giving it some numeric value. This argument is obviously useless with post-hoc simulations. Moreover, all other random processes within the CAT (such as the selection of the first items or the next item) are fixed directly within the appropriate lists (see later).

Content balancing control must be defined as input argument of randomCAT() function. The cbControl argument is a list with elements $names and $props, which are similar to those described in Sects. 4.4 and 4.6.2. Furthermore, the item bank supplied in itemBank argument must hold an additional column with subgroup membership, as illustrated in Sect. 4.2.

The nAvailable argument is used to discard some items from the set of eligible items in the bank for CAT simulation. It was primarily introduced for item exposure control purposes with the simulateRespondents() function and is not further described here.

Finally, the output of randomCAT() can be saved and stored in an external txt or csv file. To perform this storage, one has first to set the logical argument save.output to TRUE, and then to supply argument output with a vector of three character strings, specifying respectively the path for file location, the name of the file, and the type of file extension. For instance, setting this argument as c("c:/", "out", "txt") will save the output of the function call into the file out.txt and stored in the root of drive C.

Finally, the lists start, test, stop and final hold the various options for designing the CAT simulation. Together with the building and calibration of the item bank, they constitute the core of the CAT process. Each list is described hereafter in terms of available arguments, values and effects on the CAT simulation.

4.7.2 The Start List

The start list contains all information to select the first item(s) of the CAT. More precisely, the list can contain one or several of the elements listed in Table 4.4.

These elements correspond to the arguments of the startItems() function described in Sect. 4.6.2. Depending on the chosen method for selecting the first item(s), some elements may not be specified. As a reminder:

Table 4.4 Elements of the `start` list of function `randomCAT()`

Name	Function	Type	Default
fixItems	Fixes the items to administer	Numeric vector or NULL	NULL
seed	Sets the random seed to sample the starting items	Numeric or NULL	NULL
nrItems	Sets the number of items to randomly draw	Integer	1
theta	Sets the starting ability levels	Numeric vector	0
D	Sets the scaling constant D	Numeric	1
randomesque	Sets the number of *randomesque* items	Integer	1
random.seed	Fixes the seed for randomesque procedure	Numeric	NULL
startSelect	Sets the method for optimal item selection	Appropriate acronym	"MFI"

1. To let the user select the first item(s) to administer, the element `fixItems` must be specified with the item numbers in the bank. All other elements are then ignored.
2. To randomly select the first item(s), three conditions must be fulfilled:

   (a) the `fixItems` element must be set to NULL (which is its default value);
   (b) the `seed` element must be set to either a numeric value (to fix the random seed) or to NA (to allow random selection without fixing the seed);
   (c) if more than one starting item has to be selected, then the number of such items must be provided to `nrItems` element.

   All other elements are ignored.
3. If one wants to optimally select the starting item(s), it is then mandatory to keep both `fixItems` and `seed` arguments to NULL (otherwise one of the aforementioned methods will be applied). Moreover, it is mandatory to

   (a) specify the vector of starting ability levels through the `theta` element (by default it takes to the single value zero),
   (b) specify the method for selecting the starting item(s) with an appropriate value for the `startSelect` element. The default value is `"MFI"` and allowed values are `"bOpt"`, `"thOpt"`, `"progressive"` and `"proportional"` (the latter four being allowed only for dichotomous IRT models).

   Optionally, several items can be selected for each starting ability level and one of them is selected randomly as the starting item: this is set up by specifying the number of items to select with the `randomesque` element (default value is one thus no randomesque selection is performed) and the random selection can be fixed by setting a numeric value to `random.seed` element (default is NULL so random selection is not fixed).

   Thus, the method for selecting the starting item(s) has some hierarchical implementation, the setting of one method leading to ignoring all other elements that might define another approach. It is however important to notice the main exception to the hierarchy described above: If one wants to perform CAT simulations with either the proportional or the progressive approach (Barrada, Olea, Ponsoda, & Abad, 2008, 2010) throughout the whole CAT, then the only requirement is to specify the `startSelect` element to either `"progressive"` or `"proportional"`. All other arguments will be ignored (even the `fixItems` and `seed` arguments, which will be forced to take their default values in this particular case). This is because both methods for next item selection rely on the random selection of the very first item in the test.

4.7.3 The `Test` List

The `test` list holds all elements necessary to set the methods for next item selection and provisional ability estimation. They are listed in Table 4.5 together with their default values.

Table 4.5 Elements of the `test` list of function `randomCAT()`

Name	Function	Type	Default
`method`	Sets the provisional ability estimator	Appropriate acronym	`"BM"`
`priorDist`	Sets the prior distribution for ability estimation	Appropriate acronym	`"norm"`
`priorPar`	Sets the parameters of the prior distribution	Numeric vector	`c(0,1)`
`range`	Sets the range of ability estimation	Numeric vector	`c(-4,4)`
`D`	Sets the scaling constant D	Numeric	`1`
`parInt`	Sets the sequence of quadrature points	Numeric vector	`c(-4,4,33)`
`itemSelect`	Sets the method for next item selection	Appropriate acronym	`"MFI"`
`infoType`	Sets the type of item information function	Appropriate acronym	`"observed"`
`randomesque`	Sets the number of *randomesque* items	Integer	`1`
`random.seed`	Fixes the seed for randomesque procedure	Numeric	`NULL`
`AP`	Fixes the acceleration parameter for *progressive* and *proportional* methods	Numeric	`1`
`proRule`	Sets the stopping rule for *progressive* and *proportional* methods	Appropriate acronym	`"length"`
`proThr`	Sets the stopping threshold for *progressive* and *proportional* methods	Numeric acronym	`20`
`constantPatt`	sets the method to deal with constant patterns	Appropriate acronym or NULL	`NULL`

Most of these elements were described in Sect. 4.5.3. As a reminder, the four available ability estimators are *maximum likelihood* (`"ML"`), *weighted likelihood* (`"WL"`), *Bayes modal* (`"BM"`) and *expected a posteriori* (`"EAP"`). For the latter two, prior distribution and parameters are set by the elements `priorDist` and `priorPar` with the following options:

- `priorDist = "norm"` for the normal distribution and `priorPar` holds the mean and standard deviation,
- `priorDist = "unif"` for the uniform distribution and `priorPar` holds the lower and upper bound of the uniform interval,
- `priorDist = "Jeffreys"` for Jeffreys' prior and `priorPar` is ignored.

The quadrature points for EAP estimation can be defined through element `parInt`, in a regular sequence from `parInt[1]` to `parInt[2]` with length `parInt[3]`. Element D fixes the scaling constant D for logistic IRT models, and `range` sets the maximal range for ability estimates (i.e., all estimates are truncated to belong to this interval). Finally, `constantPatt` can be set to either `"fixed4"`, `"fixed8"` or `"var"`. With one of these values and in presence of a constant pattern, step-size (fixed with step .4, fixed with step .8, or variable, respectively) adjustment of the ability level is performed. In all other cases (or if `constantPatt` keeps its `NULL` default value), the provisional ability level is estimated according to the chosen `method`.

The method for next item selection is specified by the element `itemSelect`. All available methods are listed in Sect. 4.6.2. For the MEI method, the choice can be made in between using the observed or the expected information function. This can be set by the element `infoType` with possible values `"observed"` and `"expected"` respectively. The randomesque procedure to limit item overexposure can be set with the elements `randomesque` and `random.seed`, as previously described. Finally, for *proportional* and *progressive* methods, three additional elements are available:

1. the `AP` element, which determines the so-called *acceleration parameter* of both procedures (e.g., Barrada et al., 2008, 2010);
2. the `proRule` element that sets the stopping rule (which is required by both methods);
3. the `proThr` element that sets the threshold of the stopping rule (which is also required by both methods).

Note that for both compatibility and logical application of these two item selection rules, the values of elements `proRule` and `proThr` should be identical to those of the elements `rule` and `thr` of the `stop` list that is described in the next section.

Table 4.6 Elements of the stop list of function randomCAT()

Name	Function	Type	Default
rule	Sets the stopping rule(s)	Vector of appropriate acronyms	"length"
thr	Sets the threshold of the stopping rule(s)	Numeric vector	20
alpha	Specifies the significance level	Numeric	0.05

4.7.4   The Stop List

The stop list fixes the options to stop the adaptive process of a CAT. It can hold three elements, which are listed in Table 4.6 together with their default values.

At least one stopping rule must be specified through the rule element. The four possible rules are:

1. "length": the CAT stops once the required number of administered items has been reached. The thr value is the maximal test length.
2. "precision": the CAT stops when the precision on the provisional ability estimate is large enough (or equivalently, when the SE of the provisional estimate is small enough). The thr value is the minimum SE threshold.
3. "classification": also known as the *ACI rule* (Thompson, 2009), it stops the CAT when the classification threshold is not covered anymore by the provisional confidence interval (the latter being computed using provisional ability estimate, SE and significance level specified by the alpha argument). The thr value is the classification threshold.
4. "minInfo": the CAT stops when none of the available items provides more information than the specified minimum information threshold (at the provisional ability level). The thr value is the minimum information to be carried out by the available items.

Several stopping rules can be set altogether as a vector of acronyms, and then thr element must be a vector of the same length with the thresholds related to the rules set in the rule element (in the same order). The CAT will stop as soon as at least one stopping rule is satisfied. For instance, the following list

```
R> list(rule = c("length", "precision", "minInfo",
+        "classification"), thr = c(30, 0.3, 0.1, 1))
```

will lead to the end of the adaptive process when at least one of the following conditions is fulfilled:

1. a total of 30 items were administered,
2. the provisional SE of the ability estimate gets smaller than 0.3,

3. none of the available items in the bank has an information function value larger than or equal to 0.1 (at the provisional ability level),
4. the provisional 95% confidence interval of ability does not cover the classification threshold 1 anymore.

4.7.5 The Final List

The final list sets the options for final ability estimation and reporting. It holds the following elements, which are identical to those listed in Table 4.5 for the test list and take the same values and default values: method, priorDist, priorPar, range, D and parInt. Those elements define the final ability estimator, returned with the full response pattern obtained during the CAT process.

In addition, the final list can hold the element alpha in order to specify the significance level to be used for the final confidence interval of ability. The default value is 0.05 as for the stop list.

4.7.6 Output Information

The output of randomCAT() is actually a long list of class "cat" with many elements. Some of them are just recalls from the input options chosen by the user, and additional ones return the set of administered items, the corresponding item parameters, the response pattern, the provisional ability estimates and related standard errors... They can all be extracted from the output list for further use.

In addition, the print.cat() S3 function was developed to allow fancy layout of the output in the R console. It first displays a summary of the chosen CAT options, followed by a table with one column per administered item and four rows: the number in the CAT sequence (from first to last administered item), the item number, the item response, and the provisional ability estimate and related SE. The output ends up with the summary statistics: final ability estimation, SE and related confidence interval (based on the assumption of normal distribution of the ability levels). This enhanced output can be saved as an external file by specifying the arguments save.output and output as explained in Sect. 4.7.1.

Finally, it is possible to graphically display the output of randomCAT() using the corresponding S3 function plot.cat(). It basically calls the output of randomCAT() as input element x to draw the path of provisional ability estimates. Several optional arguments are listed in Table 4.7.

By default, the plot will display the trace of provisional abilities together with the true ability level. Confidence intervals can be added by setting the ci argument to TRUE. For classification rule, the classification threshold can be represented in addition by setting classThr to TRUE. Finally, options to save the plot

Table 4.7 Elements of the `plot.cat()` function

Name	Usage	Type	Default
`x`	Sets input for plotting	Object of class `"cat"`	NA
`ci`	Display confidence intervals?	Logical	FALSE
`alpha`	Specifies the significance level	Numeric	0.05
`trueTh`	Display true ability with horizontal line?	Logical	TRUE
`classThr`	Displays threshold for `"classification"` rule	Numeric or NULL	NULL
`save.plot`	Should the plot be saved?	Logical	FALSE
`save.options`	Location and format of saved plot	Vector of appropriate character strings	`c("path",` `"name",` `"pdf")`

by providing a location, a file name and extension can be provided through the arguments `save.plot` and `save.options`. Such options will be illustrated in the next chapter.

4.8   Top-Level Function: `simulateRespondents()`

The `randomCAT()` function permits designing the simulation of one CAT response pattern. In case of intensive simulation studies, it is desirable to replicate this simulation process a large number of times and under various testing conditions. One possibility is to make a repeated call to `randomCAT()` but this can lead to item exposure issues. Indeed, even though item overexposure can be limited by the various *randomesque* options to be set at the start and during the CAT, one does not have overall control to the item administration process across the whole set of test takers.

The `simulateRespondents()` function was developed to overcome this issue and provides a more global CAT generation process for many test takers. This section presents successively the input arguments and the format of the output for this function, with emphasis on both similarities and differences with `randomCAT()`.

4.8.1   Input Arguments

As it basically proceeds to repeated calls to the `randomCAT()` function, the latter function and the `simulateRespondents()` function hold many input arguments in common. They are all listed in Table 4.8, and only those arguments

Table 4.8 Input arguments for `simulateRespondents()` function

Name	Function	Type
thetas	Sets all true ability levels	Vector
itemBank	Sets the matrix of item parameters	Matrix
model	Sets the type of polytomous IRT model	Model acronym or NULL
responsesMatrix	Provides item responses for post-hoc simulations	Matrix or NULL
genSeed	Fixes the general random seed	Numeric or NULL
cbControl	Sets the options for content balancing control	Appropriate list or NULL
rMax	Sets the maximum exposure rate	Numeric
Mrmax	Sets the method to control for maximum exposure rate	Character string
start	Sets the options to select the first item(s)	Appropriate list
test	Sets the options for provisional ability estimation and next item selection	Appropriate list
stop	Sets the options of the stopping rule(s)	Appropriate list
final	Sets the options for final ability estimation	Appropriate list
save.output	Sets whether output should be saved	Logical
output	Sets the output location, file name and extension	Appropriate vector

that differ from those in `randomCAT()` (either in name, functioning, or new ones) will be described hereafter.

Arguments `itemBank`, `model`, `cbControl`, `save.output` and `output` have exactly the same meaning and functioning as in `randomCAT()` function, see Table 4.3 in Sect. 4.7.1. Moreover, the `start`, `test`, `stop` and `final` lists take exactly the same elements and values and functioning identically to those described in Sects. 4.7.2–4.7.5. Thus only five input arguments differ between the two functions and they are discussed hereafter.

The argument `thetas` in `simulateRespondents()` corresponds to the argument `trueTheta` in `randomCAT()`. However, here it may take more than one value, representing more than one response pattern to generate. In other words, `thetas` can be a vector of numeric values, each value being one particular (true) ability level to consider for CAT generation of a response pattern. Leaving a single value to `thetas` will actually force the function to make just one single call to the `randomCAT()` function and return its corresponding output described in the previous section.

Post-hoc simulations can also be performed with this function, using the argument `responsesMatrix` that acts similarly to the `responses` argument in `randomCAT()` function. However, if it is not NULL, then `responsesMatrix` must be a matrix with one row per test taker and one column per item in the bank, and the number of rows must match the length of the `thetas` argument.

Remember that the latter must be provided even with post-hoc simulations for system compatibility.

The genSeed argument has the same meaning as with randomCAT(). It must also be a vector of seed values and of the same length as vector thetas. Each run of the CAT process will be made with a different seed value to avoid generating the same response patterns and selecting the same items.

Finally, two new arguments that are specific to item exposure control are introduced. The first, argument rMax sets the maximum exposure rate per item one wishes to observe. The default value of 1 implies that each item can be selected and administered to all test takers, while smaller rates will cancel out this option. For instance, setting rMax to 0.8 means that each item in the bank cannot be administered in more than 80% of the cases to be generated, the latter being computed as the number of ability levels specified in the thetas argument.

The second new argument is called Mrmax and sets the method to constraint exposure rates to be smaller than the maximum rate specified by rMax. Possible methods are the *restricted method*, using value "restricted" (Revuelta & Ponsoda, 1998), and the *item-eligibility method*, using value "IE" (van der Linden & Veldkamp, 2004). A more detailed description of both methods can be found in Barrada, Abad, and Veldkamp (2009).

4.8.2 Output Information

The output of simulateRespondents() is a list of class "catResult" that contains many input arguments and summarizes the output from all successive calls of randomCAT() functions. The most important output elements are listed in Table 4.9.

Some elements return final results for all simulated test takers, such as the vector of final ability estimates (estimatedThetas) or a matrix with true and estimated abilities, standard errors, and test lengths (final.values.df). The full set of administered items, generated item responses, and provisional ability estimates are all available in the responses.df element. For the sake of compatibility, empty cells of this data frame (which occur when some items are not administered to some test takers) are filled in by the value −99. Moreover, all item exposure rates (i.e., proportions of item occurrences throughout the whole process) are returned through the element exposureRates.

Other summary statistics are also provided. For instance, the overall ability estimation bias (i.e., average difference between the estimated and true ability levels) and root mean squared error (i.e., square root of the average of squared differences between the estimated and true ability levels) are provided by elements bias and RMSE. The average test length is provided by the element testLength. These summary statistics are computed on the whole set of test takers, using the whole set of true ability levels. However, simulateRespondents() also returns corresponding statistics per decile. The latter is obtained as follows: the set

Table 4.9 Selected output arguments of `simulateRespondents()` function

Name	Contents
estimatedThetas	Vector of estimated ability levels
correlation	Correlation between true and estimated ability levels
bias	Bias between true and estimated ability levels
RMSE	Root mean squared error between true and estimated ability levels
exposureRates	Vector of empirical exposure rates for all items in the bank
testLength	Average test length (i.e., number of administered items)
condTheta	Vector of average ability level per decile
condBias	Vector of ability estimation bias per decile
condRMSE	Vector of RMSE values per decile
condNItems	Vector of average test length per decile
condSE	Vector of average standard error of ability per decile
overlap	Item overlap rate
final.values.df	Data frame of true ability levels and final estimates
responses.df	Data frame with all administered items and responses, and provisional ability levels
start.time	CPU time at the start of `simulateRespondents()`
finish.time	CPU time at the end of `simulateRespondents()`

of test takers is divided into ten subsets according to their true ability levels (first subset holding the smallest 10% ability levels and so on). The summary statistics (average ability estimates, bias, RMSE, average standard error of estimation, and average test length) are then computed for each and returned as vectors, in elements `condTheta`, `condBias`, `condRMSE`, `condSE` and `condNItems`, respectively.

Finally, the overall item overlap, computed as the ratio between the sum of square exposure rates and the test length, is returned by the element `overlap`, while the CPU times at the start and at the end of the whole generation process are stored in elements `start.time` and `finish.time`.

Two generic S3 functions are associated with `simulateRespondents()`. First, `print.catResult()` function displays most of this output in an optimized way for easy summary extraction of the whole process. Second, the function `plot.catResult()` graphically displays several summary statistics across all test takers or per decile. The argument `type` controls for the type of plot to be displayed. Table 4.10 lists all possible values for this argument, together with the corresponding plot.

By default, one plot with nine panels is displayed, each panel referring to one of the possible plot or scatterplot (except the one for cumulative exposure rates that is not included in this general figure). Each individual panel can be extracted using the appropriate acronym. It is also important to note that similarly to the `plot.cat()` function for `randomCAT()`, each individual plot and the general 9-panel plot can be saved as external figures or PDF files, using arguments `save.plot` and `save.options` that work identically to those described in Table 4.7.

Table 4.10 Values for type argument of plot.catResult() function

Name	Type of plot
"trueEst"	Scatterplot of true versus estimated ability levels
"expRate"	Plot of item exposure rates, sorted in decreasing order
"cumExpRate"	Plot of cumulative item exposure rates
"cumNumberItems"	Plot of length as a function of cumulative percentage of examinees (not available when stopping rule is "length")
"expRatePara"	Scatterplot of item exposure rates versus item discrimination parameters (not available when IRT model is "PCM" or "NRM")
"condBias"	Plot of conditional bias of ability estimation per true ability decile
"condRMSE"	Plot of conditional RMSE of ability estimation per true ability decile
"numberItems"	Plot of conditional test length per true ability decile (not available when stopping rule is "length")
"sError"	Plot of conditional standard error of ability estimation per true ability decile
"condThr"	Plot of conditional proportions of CATs satisfying the "precision" or "classification" stopping rule, per true ability decile (not available when stopping rule is "length")
"all"	9-panel figure of all aforementioned plots except "cumExpRate"

Chapter 5
Examples of Simulations Using catR

The previous chapter focused on the description of the R package **catR** as a tool for simulation studies on CAT processes. This chapter presents several practical illustrations of **catR** using two real item banks, one for dichotomously scored items and one for polytomously scored items. After a brief description of both item banks, several examples are displayed with the same organization.

First, the problem situation is described and the simulation plan is sketched. Then, the code to translate the problem into R execution is detailed, and the corresponding output is displayed (in part or fully, depending on the examples). This output is then further analyzed and basic conclusions are drawn for illustration purposes.

5.1 Item Banks

Two item banks calibrated under real data collections are considered in this chapter. They are briefly described and their main characteristics are outlined.

5.1.1 The Dichotomous Item Bank

The dichotomous item bank comes from a large operational assessment project. It is made up of 100 items administered to 133,132 test takers. Every test taker took the 100 items so full linear assessment was performed and a full binary matrix is available. The data were first used to create various MST designs (Yan, Lewis, & von Davier, 2014a, 2014b) and such MST structures will be considered in Chap. 8. In this chapter however, only the item bank (without modules and stages) will be used for CAT illustrations.

© Springer International Publishing AG 2017
D. Magis et al., *Computerized Adaptive and Multistage Testing with R*, Use R!,
https://doi.org/10.1007/978-3-319-69218-0_5

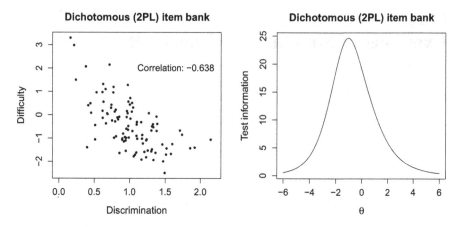

Fig. 5.1 Scatterplot of discrimination and difficulty coefficients (left) and test information function (right) of the dichotomous (2PL) item bank

Various score-based summary statistics for the dichotomous item bank are available (Yan et al., 2014b). However, to run CAT applications the items were calibrated under a 2PL parametrization using the R package **mirt** (Chalmers, 2016). Item difficulties range from −2.489 to 3.301, with a mean value of −0.414, a median value of −0.509 and a variance of 1.145. Discriminations range from 0.160 to 2.144, with a mean value of 1.008, a median value of 0.981 and variance of 0.141. The correlation between difficulties and discrimination parameters is equal to −0.638. The scatter plot of these estimated coefficients is displayed in the left panel of Fig. 5.1. Clearly, more discriminating items also have easy-to-average difficulty levels, while more difficult items are less discriminating.

It is not that usual to obtain a negative correlation between IRT difficulty and discrimination parameters. In most operational programs, more discriminating items are also more difficult. In addition, the sample correlation between proportions of correct responses (per item) and point biserial correlations (between the item responses and the total test scores) is equal to 0.415 and indicates an opposite trend. This can however be easily explained by the fact that larger proportions of correct responses correspond to easier items and thus, smaller difficulty levels. It is therefore not abnormal to observe differences between the signs of these correlations.

The right panel of Fig. 5.1 shows the overall test information function, computed as the sum of all individual information functions from all items in the bank. This information function can be obtained using **catR** function Ii (), as follows (where object it.2PL is a matrix with the item parameters from this bank):

```
R> s<-seq(-6, 6, 0.1)
```

(continued)

```
R> info.2PL<-function(t) sum(Ii(t, it.2PL)$Ii)
R> TIF.2PL<-sapply(s, info.2PL)
R> plot(s, TIF.2PL, type="l",
+        xlab = expression(theta),
+        ylab = "Test information")
```

The item bank is most informative around the center of the scale, where most of the discriminative items are located, while at the extremes of the scale fewer items are present and these items are also less informative. One can therefore expect better efficiency of estimations of ability around the center of the ability scale. This bank is further referred to as the *2PL item bank*.

5.1.2   The Polytomous Item Bank

The polytomous item bank is called the CAT-PAV questionnaire (Santos, 2017) and is a collection of 96 open-ended items that was developed to propose a computer-adaptive assessment of productive and contextualized academic English vocabulary, piloted with students at Iowa State University (USA). Items are presented as two sentences drawn from the academic subset of the COCA corpus, with one common missing word to be identified by test takers. In case of an incorrect first answer, synonyms are exhibited as a hint. Th item is scored 2 if a correct answer was provided from first attempt, 1 if a correct answer was provided after the hint being shown, and 0 in case of an incorrect answer after the hint was shown.

Items were field tested at Iowa State University and calibrated in three batches. The first batch (with the first 38 items) was administered to 140 test takers, the second batch (with items 39–68) to 153 test takers, and the third batch (with items 69–96) to 110 test takers. Several items from batch 1 were also included to act as anchor items for equating purposes. Each batch was calibrated separately under the GPCM model (2.15) using software **Jmetrik** (Meyer, 2014, 2015). The overall test information function for this polytomous item bank can be obtained using the following code, where it.GPCM holds the bank item parameters:

```
R> s<-seq(-6, 6, 0.1)
R> info.GPCM<-function(t) sum(Ii(t, it.GPCM)$Ii)
R> TIF.GPCM<-sapply(s, info.GPCM)
R> plot(s, TIF.GPCM, type="l",
+        xlab = expression(theta),
+        ylab = "Test information")
```

Fig. 5.2 Test information
function of the polytomous
(GPCM) item bank

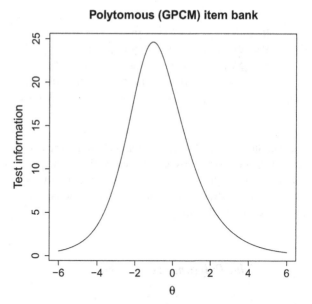

Corresponding graph is displayed in Fig. 5.2. This polytomous item bank (further referred to as the *GPCM item bank*) does not exhibit uniform information along the ability scale. It is most informative slightly below the center of the scale and is almost non informative at the extremes of the scale. Further details about the item bank construction, calibration and evaluation are available in Santos (2017).

5.2 Example 1a: CAT with Dichotomous Item Bank

To get started with **catR**, we provide a first illustration of the application of the randomCAT() function. The following design is considered.

One response pattern is generated for a test taker whose true ability level is 0.5. The CAT is initialized by selecting the most informative item at ability level zero. Within the test step, the next item is chosen by the maximum Fisher information (MFI) criterion and provisional ability estimation is performed with Bayes modal method (and by-default standard normal prior distribution). The test stops when its length reaches 20 items. Final ability estimation is performed by BM and standard normal prior distribution (i.e., the same estimator is used throughout the CAT). The 2PL item bank is used for this example.

This scenario can be built with the following R code to set up the four lists and the call to randomCAT() function. Note that the random generation is fixed by using genSeed = 1 so that all forthcoming results can be replicated with the same code.

```
R> start <- list(theta = 0, startSelect = "MFI")
R> test <- list(method = "BM", itemSelect = "MFI")
R> stop <- list(rule = "length", thr = 20)
R> final <- list(method = "BM")
R> ex1 <- randomCAT(trueTheta = 0.5,
+                   itemBank = it.2PL, start = start,
+                   test = test, stop = stop,
+                   final = final, genSeed = 1)
```

With this first illustration, the full output is displayed below.

```
R> ex1
Random generation of a CAT response pattern
  with random seed equal to 1

  Item bank calibrated under Two-Parameter Logistic
  model

  True ability level: 0.5

  Starting parameters:
    Number of early items: 1
    Early item selection: maximum informative item
      for starting ability
      Early items not chosen to control for content
        balancing
    Starting ability: 0

  Adaptive test parameters:
    Next item selection method: maximum Fisher
      information
    Provisional ability estimator: Bayes modal (MAP)
      estimator
      Provisional prior ability distribution: N(0,1)
        prior
    Ability estimation adjustment for constant
      pattern: none

  Stopping rule:
```

(continued)

```
     Stopping criterion: length of test
      Maximum test length: 20

  Randomesque method:
    Number of 'randomesque' starting items: 1
    Number of 'randomesque' test items: 1

  Content balancing control:
    No control for content balancing

  Adaptive test details:

Nr           1      2      3      4      5      6      7
Item        12     87     83     46     74     15     88
Resp.        1      1      1      0      1      1      1
Est.     0.282  0.579  0.803  0.536  0.652  0.748  0.889
SE       0.844  0.775  0.732  0.655  0.628  0.609    0.6

Nr           8      9     10     11     12     13     14
Item        79     47     35     43     91     94     40
Resp.        0      1      0      1      1      0      1
Est.     0.715  0.842   0.69  0.748  0.835  0.747  0.804
.SE      0.554  0.548  0.515  0.504  0.498  0.477   0.47

Nr          15     16     17     18     19     20
Item        37     32      9     90     36     11
Resp.        1      1      1      1      1      1
Est.     0.845  0.895   0.96  1.018  1.065  1.111
SE       0.463  0.458  0.455  0.451  0.448  0.446

  Satisfied stopping rule:
    Length of test

  Final results:
    Length of adaptive test: 20 items
    Final ability estimator: Bayes modal (MAP)
                              estimator
    Final prior distribution: N(0,1) prior
    Final ability estimate (SE): 1.111 (0.446)
    95% confidence interval: [0.238,1.985]

Output was not captured!
```

The first part of the output summarizes the various input options selected through the four input lists. By-default option values are also mentioned for completeness: for instance, there was no control for content balancing and no randomesque selection of the items. The function also detected that the item bank was calibrated under the 2PL model (as a consequence of the item bank coding within it.2PL matrix).

The second half of the output returns the adaptive test details and the final results from CAT generation. First, an array with five rows and as many columns as the number of administered items is displayed. Each column corresponds to one administered item and the five rows return the following information: Nr is the item number in the CAT process (i.e., first administered item, second administered item etc.); Item is the item identification number in the bank; Resp is the item response generated by **catR**; Est is the provisional ability estimate after the item was administered and response generated; and SE is the corresponding provisional SE value. In this example, 20 items were administered, as expected from the CAT design. Note that SE values are decreasing when the test length increases (from 0.844 after the first item to 0.446 at the end of the CAT), which was also expected.

The last part of the output displays summary measures at the end of the CAT. The length of the CAT and the chosen final ability estimator are mentioned first, followed by the final ability estimate, related SE and corresponding 95% confidence interval. Note that the latter correctly covers the true ability level of 0.5. Also, final results are identical to the values obtained after the 20th item administration, which is obvious since provisional and final ability estimators are identical.

The whole trajectory of ability estimates, from the start to the end of the CAT, can be graphically displayed using the following code (note that option ci = TRUE was added to simultaneously represent the sequence of confidence intervals at each step of the CAT):

```
R> plot(ex1, ci = TRUE)
```

Figure 5.3 holds the corresponding plot. The horizontal line represents the true ability level. This figure illustrates that throughout the CAT the ability estimate has a global increasing trend, departing somewhat from the true level. Yet, all confidence intervals accurately cover the true ability level, even while the range of such intervals decreases with the test length (as a consequence of related SE reduction).

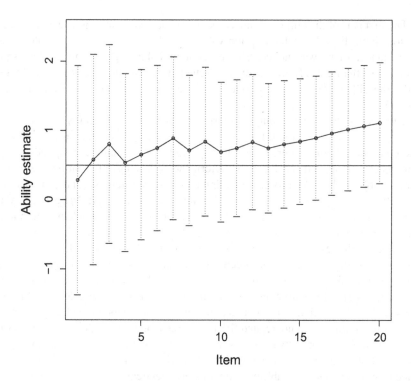

Fig. 5.3 Sequences of successive ability estimates and related 95% confidence intervals with the 2PL item bank

5.3   Example 1b: CAT with Polytomous Item Bank

The second illustration is similar to the previous one, except that (a) the GPCM item bank will be considered instead, and (b) some CAT options will be modified for the sake of illustration. The design is chosen as follows.

One response pattern is generated for a test taker with true ability level equal to −1. The CAT is initialized by selecting two items at random. In the test step, the next item is chosen by maximum expected information (MEI) criterion and provisional ability estimation is performed by weighted likelihood (WL) method. The test stops when either its length reaches 30 items or when the provisional SE becomes less than or equal to 0.25. Final ability estimation is performed by ML.

The following R code sets up this scenario and perform related computations. Random response generation is fixed by using `genSeed = 1` and random starting item selection with `seed = 1`, so that all random generations can be reproduced.

```
R> start <- list(nrItems = 2, seed = 1)
R> test <- list(method = "WL", itemSelect = "MEI")
R> stop <- list(rule = c("length", "precision"),
+                thr = c(30, 0.25))
R> final <- list(method = "ML")
R> ex2 <- randomCAT(trueTheta = -1,
+          itemBank = it.GPCM, model = "GPCM",
+          start = start, test = test, stop = stop,
+          final = final, genSeed = 1)
```

Selected pieces of the output are displayed below.

```
R> ex2
Random generation of a CAT response pattern
  with random seed equal to 1

  Item bank calibrated under Generalized Partial
    Credit Model

  True ability level: -1

  Starting parameters:
    Number of early items: 2
    Early items selection: Random selection in
     item bank
     Early items not chosen to control for content
      balancing
    Items administered: 26 and 36

  Adaptive test parameters:
    Next item selection method: Maximum expected
     information (MEI)
    Provisional ability estimator: Weighted
     likelihood estimator

  (...)

  Stopping rules:
    Stopping criterion 1: length of test
```

(continued)

```
       Maximum test length: 30
       Stopping criterion 2: precision of ability
         estimate
       Maximum SE value: 0.25

   (...)

    Adaptive test details:


   Nr          1        2        3        4        5        6
   Item       26       36       25       67       60       15
   Resp.       2        1        1        0        0        2
   Est.       NA  -0.643  -0.617  -0.741  -0.883  -0.757
   SE         NA   0.833      0.6   0.508   0.462   0.414


   Nr          7        8        9       10       11       12
   Item       58       19       12       27       17       35
   Resp.       0        1        0        2        2        0
   Est.   -0.879  -0.863  -0.925  -0.871  -0.802  -0.861
   SE       0.39   0.365     0.35   0.329   0.313   0.301


   Nr         13       14       15       16       17       18
   Item       90       68       41       16       65        8
   Resp.       1        0        0        0        0        2
   Est.   -0.853  -0.918  -0.967  -1.018   -1.05  -0.994
   SE       0.29   0.283   0.277   0.273   0.268   0.259


   Nr         19       20
   Item       54       62
   Resp.       0        2
   Est.   -1.018  -0.967
   SE      0.255   0.247


    Satisfied stopping rule:
      Precision of ability estimate

    Final results:
      Length of adaptive test: 20 items
      Final ability estimator: Maximum likelihood
        estimator
      Final range of ability values: [-4,4]
```

(continued)

```
Final ability estimate (SE): -0.98 (0.248)
95% confidence interval: [-1.466,-0.494]
```

In this example, the CAT process stopped after 20 administered items since the provisional SE value (0.248) became less than the stopping threshold (0.25). The final ML estimate and SE values are very close to the provisional ones at the end of the CAT, though computed with the WL estimator. The final 95% confidence interval actually covers the true ability level used to generate the CAT pattern.

5.4 Example 2: CAT for Placement Tests

One of the main assets of CAT is that it can significantly decrease the length of the test without affecting the quality of the estimated abilities too much. The purpose of this example is to illustrate this phenomenon by comparing linear and CAT placement tests with the same set of simulated data. More precisely, classifications of test takers with respect to one specified "pass-fail" threshold will be compared, both with the full 2PL item bank (linear test) and with CAT design (using post-hoc simulations) with at most one half of the bank being administered.

5.4.1 Data Generation and Linear Design

The generated data set follows the true underlying 2PL bank structure. The true ability levels are taken from a regular sequence form -2 to 2 by steps of 0.2 (so 21 different values in total) and 1000 response patterns are generated for each true ability level, yielding thus 21,000 simulated respondents. This will allow for comparison of linear and CAT efficiencies in classification for a fine grid of true ability levels and with enough test takers at each ability level.

The data generation process can be performed as follows, using **catR** function `genPattern()`:

```
R> s <- seq(-2, 2, 0.2)
R> th <- rep(s, each = 1000)
R> data <- genPattern(th, it.2PL, seed = 1)
```

Now, the `thEst()` function is created to extract, for a given response pattern from the `data` set, the lower and upper bounds of the 95% confidence interval computed under ML estimation of ability and the whole item bank (100 items in total), using functions thetaEst() and semTheta():

```
R> thEst <- function(x) {
+         pr <- thetaEst(it.2PL, x, method = "ML")
+         se <- semTheta(pr, it.2PL, x,
                  method = "ML")
+         res<-c(pr+qnorm(.025)*se,
                  pr+qnorm(.975)*se)
+         return(res) }
```

Extracting lower and upper bounds for each generated response pattern can be done using the following code (returning a matrix with 21,000 rows and two columns, one row per test taker and the two confidence bounds in each column):

```
res.linear <- t(apply(data, 1, thEst))
```

This output from linear (simulated) assessment will be further compared to the corresponding CAT output.

5.4.2 CAT Design and Implementation

The following CAT design is considered in order to compare CAT and linear classification performances. First of all, post-hoc simulations using the data set were run to compare the results with the same item responses for the test takers, and the classification threshold of -1.048 was selected as it corresponds to the ability level that maximizes the bank information function (as illustrated in right panel of Fig. 5.1). Second, specific CAT options were fixed as follows. Each CAT starts with two items being chosen as most informative around ability values of -2 and 0 (that is, almost evenly spread around the classification threshold), with randomesque selection among the five optimum items per starting ability value. The next item is chosen using Kullback-Leibler (KL) information. Provisional and final ability estimation is performed by ML method (as in linear testing), with variable stepsize adjustment at the start of the CAT in case of a constant pattern. Eventually, the CAT stops when either the test taker can be classified with respect to threshold -1.048 with 95% confidence, or when the test reaches 50 items (that is half the size of the total item bank). The extracted output for each generated CAT consists of the total test length (with a maximum of 50), and the lower and the upper bounds of the final 95% confidence interval.

This design can be implemented as follows. To fix the seed of randomesque selection, the definition of starting and test lists are embedded into a loop over all test takers. The output is stored into the `res.cat` matrix with one row per test taker and three columns (test length, lower bound and upper bound respectively). Note that although post-hoc simulations are specified by the `responses` argument, some `trueTheta` input value must be provided for compatibility. Here, this input value is set to zero but is not used further in the process.

```
R> res.cat <- matrix(NA, nrow(data), 3)
R> for (i in 1:nrow(data)){
+   start <- list(theta = c(-2, 0), randomesque = 5,
+     random.seed = i)
+   test <- list(method = "ML",
                      constantPatt = "var",
+     itemSelect = "KL", randomesque = 10,
+     random.seed = i)
+   stop <- list(rule = c("length",
+     "classification"), thr = c(50, -1.048))
+   final <- list(method = "ML")
+   pr <- randomCAT(trueTheta = 0,
                      itemBank = it.2PL,
+     responses = data[i,], start = start,
+     test = test, stop = stop, final = final)
+   res.cat[i,] <- c(length(pr$pattern), pr$ciFinal)
+ }
```

5.4.3 Output Analysis

The two output matrices `res.linear` and `res.cat` can now be further analyzed together with the vector `th` of true ability levels used to generate the data. The full R code is not displayed here since it relies on basic use and implementation in the R language and does not relate to package **catR**. An output CSV file with the true ability levels, confidence bounds for linear and CAT assessments, and CAT test lengths, is available upon request.

For linear and CAT assessments, three summary statistics were computed per true ability level: the proportion of correct classifications (i.e., proportions of test takers whose confidence interval lies on the correct side of the classification threshold), the proportion of incorrect classifications (i.e., proportions of test takers whose confidence interval lies on the incorrect side of the classification threshold), and the proportion of undetermined classifications (i.e., proportions of test takers whose confidence interval overlaps with the classification threshold). These three curves

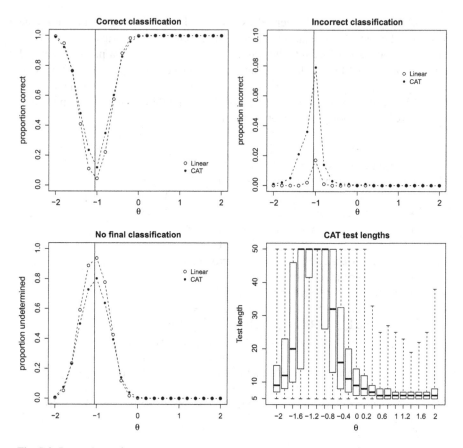

Fig. 5.4 Proportions of test taker correctly classified (upper left), incorrectly classified (upper right) and without final classification (lower left) for linear and CAT assessment based on post-hoc simulations with 2PL item bank. Boxplots of CAT test lengths (lower right) are displayed for each true ability level. Classification threshold (−1.048) is displayed by a vertical solid line

are displayed in upper left, upper right, and lower left panels of Fig. 5.4, respectively. In addition, boxplots of the distributions of CAT test lengths per true ability level are displayed on the lower right panel of this Figure.

First of all, almost all results displayed in Fig. 5.4 are relatively symmetric around the value −1, which can be expected from the shape of the bank information function displayed in Fig. 5.1, right panel. Second, differences between linear and CAT assessments mostly occur around the classification threshold. At both extremes of the ability scale (and more striking at the upper limit of this scale), the proportions of correct classifications are almost maximum and so the other two types of proportions are close to zero. Even though informative items are less available in these areas of ability, the gap between the classification threshold and the true ability level is large enough to warrant correct classification of the test takers, based on their final 95% confidence interval.

As the ability level gets closer to the classification threshold, expected trends also appear: the proportion of correct classifications decreases dramatically while the proportion of undetermined classifications increases in an almost compensatory way. In addition, the proportion of incorrect classifications also increases but to a lower extent (less than 8% in all cases). This translates the fact that discriminating the true ability levels around the classification threshold is much harder than for ability levels that are far away from the threshold. For instance, for true ability level -1, about 80% of test takers do not reach final classification. The proportion of incorrect classifications also raises slightly around that threshold, indicating the increased risk of making a classification mistake when one is located close-by to the decision threshold.

Eventually, interesting differences occur between the assessment modes around the classification threshold. First, the proportion of correct classifications is larger for CAT design (11.9% for ability level -1) than for linear design (4.5% at the same level). Second, the proportion of incorrect classifications is larger for CAT than for linear testing: 7.9% versus 1.7% at ability level of -1. Eventually, the proportions of undetermined classifications are larger for linear testing (maximum value of 93.8%) than for CAT (corresponding proportion of 80.2%). Altogether, these results indicate that although administering fewer items than in linear testing, CAT designs return more often some final classification (or equivalently fewer undetermined conclusions), which leads to an increase in both proportions of correct and incorrect classifications. However, among undetermined cases with linear testing that yield a classification with CAT, there are globally more correct than incorrect classifications. This can also be observed from Table 5.1 which displays the cross-tabulation of test takers with the three possible final classification outcomes (incorrect, undetermined, correct) for both linear and CAT designs.

In sum, CAT design improved or maintained the classification of most of the test takers with a test length reduction of at least 50%, even more when the true ability level lies far away from the classification threshold (as can be seen from the lower right panel of Fig. 5.4). CAT returns final classifications more often than linear testing, at the cost of an increased incorrect classification rate around the

Table 5.1 Contingency table of number of final classification outcomes (incorrect, undetermined, correct) for both linear and CAT designs

| Linear | CAT | | | |
	Incorrect	Undetermined	Correct	Total
Incorrect	16	4	0	20
Undetermined	143	3067	838	4048
Correct	5	489	16,438	16,932
Total	163	3560	17,276	21,000

threshold. But the gain looks to be globally significant overall. Examples of further analysis could include, for instance, different stopping rules with longer or shorter tests, different rules to select the next item or to estimate the ability levels.

5.5   Example 3: CAT with Unsuitable Data

An interesting question is to investigate the efficiency in ability estimation with CAT post-hoc simulations using data that do not follow the underlying item bank structure. This may be a somewhat common problem in CAT, as well-calibrated item banks could be too often used for further CAT administrations while violating the latent structure (i.e., true underlying IRT model) of the data. One way to analyze this potential problem is to run simulations with **catR**. Two steps are mandatory: the generation of "true" data (using the true item bank) and "corrupted data" (with respect to some modified item bank), and the comparison of CAT administrations with both data sets in terms of global estimation performances. For benchmarking reasons CAT estimates will also be compared to linear estimates on the whole set of items.

5.5.1   Data Generation

Two data sets are generated for the same set of test takers. True ability levels were selected from the regular sequence of values from -2 to 2 by steps of one half. For each true level, 1000 test takers were considered, leading to a total of 9000 test takers. This choice allows comparing the two types of data (and underlying true structures) for several ability levels and with enough replicated patterns to allow some high precision in the final results.

The first data set is the "true" one, for which item responses are generated according to the parameters of the 2PL item bank. The second data set is the "corrupted" one and patterns are generated by using a modified version of the 2PL bank, which is called the 4PL item bank. More precisely, items in the 4PL bank have exactly the same difficulty and discrimination levels as items from the 2PL bank, but all lower asymptotes are now drawn from a uniform distribution between 0 and 0.25. Moreover, the first half of the items in the bank also have upper asymptote parameters drawn from a uniform distribution on the range [0.95; 1]. In this way, the 4PL bank has the same number of items and the same difficulty and discrimination values as the 2PL bank, but half of the items are now generated according to a 3PL model and the other half from a 4PL model.

The following R code can be used and the data are easily generated with the `genPattern()` function from **catR**. Random seed was fixed to allow replicating the results.

```
R> set.seed(1)
R> it.4PL <- it.2PL
R> it.4PL[, 3] <- runif(100, 0, 0.25)
R> it.4PL[1:50, 4] <- runif(50, 0.95, 1)
R> s <- seq(-2, 2, 0.5)
R> th <- rep(s, each = 1000)
R> data2PL <- genPattern(th, it.2PL, seed = 1)
R> data4PL <- genPattern(th, it.4PL, seed = 1)
```

5.5.2 CAT Design and Implementation

The following CAT options are considered, both with the 2PL and the 4PL data sets
to make comparable designs. At the starting step, one item is chosen with difficulty
level as close as possible to starting ability zero (the thOpt rule), and randomesque
selection among the top ten items is performed. Within the CAT, the next item is
chosen by MLWI rule and ability is estimated by EAP with standard normal prior
distribution. Randomesque selection of the next item is also performed among the
ten best candidates. Only one stopping rule is considered, that is, when the test
reaches 30 items (this allows further comparisons between the two data sets at fixed
test length). Final EAP ability estimates are computed and stored as final output.
To allow replications and fix the random seed values accurately, the following loop
was designed in R. Vectors th.2PL and th.4PL hold final CAT ability estimates
with both data sets.

```
R> th.2PL <- th.4PL <- NULL
R> for (i in 1:length(th)){
+ start <- list(theta = 0, startSelect = "bOpt",
+               randomesque = 10, random.seed = i)
+ test <- list(itemSelect = "MLWI", method = "EAP",
+               randomesque = 10, random.seed = i)
+ stop <- list(rule = c("length"), thr = 30)
+ final <- list(method = "EAP")
+ th.2PL[i] <- randomCAT(trueTheta = 0,
+               itemBank = it.2PL,
+               responses = data2PL[i,],
+               start = start, test = test,
+               stop = stop, final = final)$thFinal
+ th.4PL[i] <- randomCAT(trueTheta = 0,
```

(continued)

```
+                        itemBank = it.4PL,
+                        responses = data4PL[i,],
+                        start = start, test = test,
+                        stop = stop, final = final)$thFinal
}
```

In addition, linear ability estimates based on the whole set of 100 items can be obtained with the thetaEst() function, using the following code.

```
R> th.2PL.lin <- th.4PL.lin <- NULL
R> for (i in 1:length(th)){
+ th.2PL.lin[i] <- thetaEst(it.2PL, data2PL[i,],
+                           method = "EAP")
+ th.4PL.lin[i] <- thetaEst(it.4PL, data4PL[i,],
+                           method = "EAP")
+}
```

Both CAT and linear ability estimates are available in a CSV file upon request.

5.5.3 Results

For each data set and each design (CAT versus linear), averaged signed bias (ASB) and root mean squared error (RMSE) values were computed at each true ability level. ASB is the mean difference between estimated and true ability levels, while RMSE is the square root of the mean of the squared differences between estimated and true ability levels. These summary statistics are displayed in Fig. 5.5, for CAT (circles) and linear (triangles) designs and for both the "true" 2PL data (empty symbols) and the "corrupted" 4PL data (full symbols).

Linear testing returns overall better results (lower ASB and RMSE) than CAT. This was expected as linear testing involves 100 items while CATs are based on tests of 30 items only. Note however that the gap between linear and CAT designs is not that large, which indicates that the CAT design can accurately estimate ability with some limited estimation bias and loss of precision.

Moreover, both data sets display the same trends in the final results. Ability is slightly overestimated for low-ability test takers while it gets largely underestimated for higher ability levels. This is due to the shrinkage effect from the use of the EAP ability estimator with standard normal prior distribution. Consequently, larger RMSE values at the extremes of the ability scale. Note that for both ASB

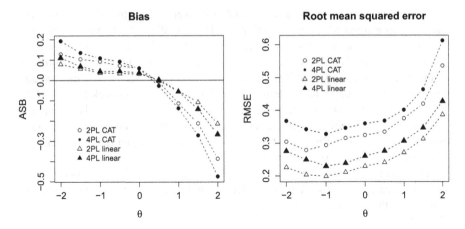

Fig. 5.5 Bias (left) and root mean squared error (right) curves of EAP ability estimator for the 2PL (true) data and the 4PL (corrupted) data sets

and RMSE there is an apparent asymmetry around ability level zero (larger bias and RMSE for more extreme ability levels), which is also expected since the 2PL item bank has maximum information around ability level -1 and the prior distribution is centered around zero.

Essentially, results with the 2PL data are more accurate than those with 4PL data: smaller ASB (in absolute value) and smaller RMSE overall. For bias values, the gap between the two data sets is slightly larger in CAT designs than in linear designs, which is due to the difference in test lengths. For RMSE values, such a gap between both designs is almost absent.

This example illustrates the potential failure when administering a CAT (or linear) test to test takers whose response trends do not follow the underlying IRT structure. In this case, with sufficiently long CATs, the gap between true and corrupted data remains quite small and acceptable, but other CAT designs (i.e., different stopping rules, different item selection rules...) could reveal stronger weaknesses.

5.6 Example 4: simulateRespondents() Function

To end this chapter we will illustrate the simulateRespondents() function using the 2PL item bank. As explained in Chap. 4, this functions allows repeated CAT pattern generation for a large number of test takers, under regular or post-hoc simulations. In other words, examples described in Sects. 5.4 and 5.5 could also be conducted with simulateRespondents() but the latter offers some additional options and automatic summary output. The design is first described, followed by the corresponding R code and the full output is displayed afterward.

The simulation focuses on a set of 1000 test takers whose true ability levels are drawn from a standard normal distribution. The overall CAT design is configured for each test taker as follows. First, one starting item is chosen as most informative for starting ability level zero, with randomesque selection among the 10 optimum items. Second, provisional ability estimation is obtained with the WL method and the next item selection is performed by maximizing the Kullback-Leibler (KL) information, using the randomesque procedure with 10 items too. Two stopping rules are specified: either when the test reaches 45 administered items or when the SE of the estimated ability level becomes smaller than or equal to 0.3. Eventually, final ability estimation is performed by WL.

The generation of true ability levels and the design of the CAT options can be implemented in **catR** as follows. The random seed value is fixed to one for reproducibility reasons.

```
R> set.seed(1)
R> TH <- rnorm(1000)
R> start <- list(theta = 0, startSelect="MFI",
+                randomesque = 10)
R> test <- list(method = "WL", itemSelect = "KL",
+               randomesque = 10)
R> stop <- list(rule =c("length", "precision"),
+               thr = c(45, 0.3))
R> final <- list(method = "WL")
```

Now, despite an important control of item over-exposure, thanks to the randomesque selection of starting and next items, an overall maximum exposure rate of 60% will be set as a global constraint. That is, the same item cannot be administered to more than 600 test takers (among the 1000 generated ones). The by-default restricted method for item exposure control is selected. Moreover, recall that the seed for random generation of response patterns can be fixed by a single input argument genSeed in the simulateRespondents(), which is an asset to previous examples that required repeated calls to randomCAT() function through a loop. The following code executes the function with aforementioned settings.

```
R> res.ex4 <- simulateRespondents(thetas = TH,
+            itemBank = it.2PL, genSeed = 1:1000,
+            rmax = 0.6, start = start, test = test,
+            stop = stop, final = final)
```

The full output is displayed below. Several interesting summary statistics (such as the total computing time, the average test length, the minimum and maximum observed exposure rates, among others) are displayed, together with usual recalls of CAT options (for instance the next item selection and the stopping rules). In this illustration, about 24 min were needed to obtain the final output (this value obviously depends on both the chosen CAT options and the availability of powerful computer resources). The average test length (across all test takers) is a bit less than 40 items while the maximum test length was fixed to 50 (compared to the 100 items in the bank). Computed across all test takers, the correlation between true and estimated ability levels is 0.603, the bias equals 0.386 and the root mean squared error is equal to 1.653. Such summary statistics provide evidence about the overall efficiency of the CAT process and can be used to compare various CAT designs and options using the same item bank and set of simulated test takers.

```
R> res.ex4

** Simulation of multiple examinees **

Random seed was fixed (see argument 'genSeed')

Simulation time: 23.4987 minutes

Number of simulees: 1000
Item bank size: 100 items
IRT model: Two-Parameter Logistic model

Item selection criterion: KL
 Stopping rules:
         Stopping criterion 1: length of test
             Maximum test length: 45
         Stopping criterion 2: precision of ability
    estimate
             Maximum SE value: 0.3
  rmax: 0.6
             Restriction method: restricted

Mean test length: 39.834 items
Correlation(true thetas,estimated thetas): 0.603
RMSE: 1.6528
Bias: 0.3855
Maximum exposure rate: 0.6
Number of item(s) with maximum exposure rate: 14
```

(continued)

```
Minimum exposure rate: 0
Number of item(s) with minimum exposure rate: 9
Item overlap rate: 0.5037

Conditional results
                        Measure     D1       D2       D3
                     Mean Theta  -1.85   -1.085   -0.695
                           RMSE  2.184     1.58    1.379
                      Mean bias  0.424    0.603    0.509
               Mean test length  36.79     30.3    36.22
          Mean standard error  0.452    0.357    0.348
  Proportion stop rule satisfied      1        1        1
              Number of simulees    100      100      100
       D4      D5      D6      D7      D8      D9     D10
   -0.391  -0.153   0.105   0.402   0.688   1.052    1.81
    1.21    1.195   1.669   1.289   1.692   2.009   1.978
   -0.197  -0.067  -0.377   0.231   0.828   0.962    0.94
    39.98   41.04   42.17   41.71   43.04   42.91   44.18
    0.331   0.344   0.381   0.394   0.508   0.616   0.768
        1       1       1       1       1       1       1
      100     100     100     100     100     100     100

These results can be saved by setting 'save.output'
to TRUE in the 'simulateRespondents' function
```

The output also returns statistics related to the item exposure rate. Maximum exposure was fixed to a rate of 0.6 and this rate was reached for 14 administered items. On the other hand, minimum observed exposure rate was 0 and was observed for 9 items. That is, 9 out of the 100 items in the bank have not been administered in this full design.

Finally, the output provides summary statistics for test takers subdivided into ten subgroups of equal (or approximately equal) sizes by estimated ability levels. The final table of the output displays, for each subgroup of test takers: the average true ability level (Mean Theta), RMSE and bias values, average test length and standard errors, as well as the proportion of test takers for which the stopping rule was satisfied. Note that in this illustration these proportions are all equal to one because of the inclusion of the length criterion.

Several elements of this output can be graphically displayed. By default the following basic plot code:

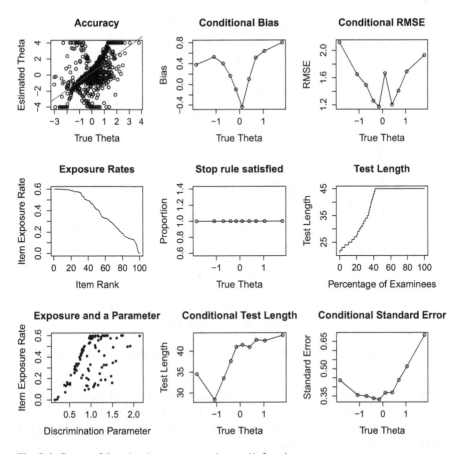

Fig. 5.6 Output of the simulateRespondents() function

```
R> plot(res.ex4)
```

returns a set of 9 panels displayed in a 3 × 3 matrix format, as shown in Fig. 5.6. From left to right and top to bottom, the panels display: (1) the scatter plot of true versus estimated ability levels; (2) the conditional estimation bias ("conditional" with respect to the ten subgroups of test takers); (3) the conditional RMSE values; (4) the trace plot of exposure rates, sorted in decreasing order according to the item exposure rank; (5) the conditional proportions of cases with satisfied stopping rules; (6) the trace plot of test lengths, sorted in increasing order; (7) the scatter plot of item exposure rate versus item discrimination parameter (more discriminative items being more informative and thus usually more often selected and exposed); (8) the conditional average test lengths; and (9) the conditional average standard errors.

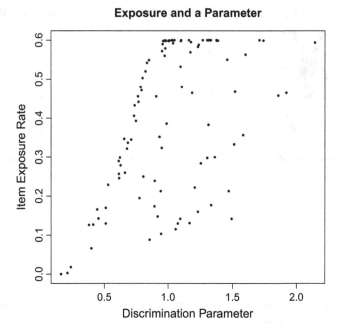

Fig. 5.7 Output of the "expRatePara" panel from simulateRespondents() function

Note that each panel of Fig. 5.6 can be extracted individually using the same function and adding an appropriate value to the type argument. For instance, type = "expRatePara" will display the scatter plot of item discrimination levels versus item exposure rates:

```
R> plot(res.ex4, type = "expRatePara")
```

which returns the following single panel:

The list of all possible values for the type argument and the corresponding plots is available in Table 4.10 (Fig. 5.7).

Part II
Computerized Multistage Testing

Chapter 6
An Overview of Computerized Multistage Testing

In this chapter, we present a brief overview of computerized multistage testing theory, including test design, test assembly, item bank, module selection, routing, scoring and linking, and exposure and security. We also provide a summary of the IRT-based module selection process, as well as the tree-based multistage testing.

6.1 Introduction and Background

Similarly to computerized adaptive testing (CAT), multistage testing (MST) is a test design in which the difficulty of the test is adapted to the level of ability of a test taker during the test administration. The MST can be seen as a hybrid of a CAT and a linear test and includes features from both designs. MST became of increasing interest in the testing industry as technology advanced and allowed for a fine-tuned engineering of the MST implementation.

The first multistage tests emerged in 1950s and 1960s and were designed in the classical test theory framework (Angoff & Huddleston, 1958; Cronbach & Gleser, 1965; Linn, Rock, & Cleary, 1968). Multistage adaptive tests were developed to tailor the test to the students' ability. Initially, the multistage tests were used more often for classification rather than for ranking the students (Smith & Lewis, 2014). The work on the tailored tests in an IRT context began with Lord (1971b) and Weiss (1983). Lord's overview (Lord, 1980) of two-stage testing set a rich research agenda for the years to come.

MST and CAT are similar in the sense that the sequence in which the items on the test are administered to test takers depends on the test takers' performance on the previous items. However, in an MST, the algorithm adapts after the test taker responds to a set of items (called modules) instead of after each item as in a CAT. Specifically, the test administration begins by administering a group of items, called the routing module, to all test takers. Then, each test taker's performance on the

© Springer International Publishing AG 2017
D. Magis et al., *Computerized Adaptive and Multistage Testing with R*, Use R!,
https://doi.org/10.1007/978-3-319-69218-0_6

items in the routing module is calculated and compared to a criterion score. A test taker is next administered a more difficult set of items (a more difficult module) if the score was higher than the criterion score, or an easier module if the test taker's score was lower than the criterion score.

Hence, an MST is a compromise between a linear test and a CAT in terms of flexibility, complexity, and practical potential. An MST is shorter in test length than a linear test and it can be almost as efficient as a CAT with respect to the measurement of the test takers' ability. The more (short) stages exist in an MST, the more it will resemble a CAT.

In this chapter, we provide an overview of the MST design using the notations and methodology introduced in the earlier chapters. We start by describing the MST in detail, then discuss the design considerations, and the methods used for calibration and scoring (IRT and CART, as in the case of CAT).

6.2   MST Basics

A multistage adaptive test is a test composed of pre-assembled short linear tests called *modules* and is administered in stages (minimum of 2). These modules have different levels of difficulty and meet the test requirements about item exposure and content representation. The adaptive part of the test is attained by selecting a module for a test taker at each stage according to his/her performance on previous stages (example of MST design in Fig. 6.1). This process reduces the test length without necessarily losing much information, because the items target specific ability levels more precisely than a linear test. The modules in the first stage (usually, there is only one module in stage 1) are called *routing modules* and the module selection criteria is called *Routing*. Some manual or semi-manual test assembly procedures allow for a revision of the content of each module by the test developers and it also facilitates the implementation of item revision within each module (Yan, von Davier, & Lewis, 2014). As it will be discussed later, this advantage may be lost if the test assembly is conducted automatically such as in the shadow testing method (van der Linden & Diao, 2014).

In Fig. 6.1 we illustrate an MST design with three stages, with one module in the first stage (the routing module), two modules (of low- and high-difficulty level) in the second stage, and with three modules at stage three (of easy, medium, and high difficulty). A test taker will first be presented with the routing module and then, based on his/her performance on these items, he/she will be routed to an easier or more difficult module on the next stage. This principle will be applied for the routing to the next stage(s).

The structure compounded by the routing stage, subsequent stages and modules is called a *panel*. In a standardized, operational test, a multistage test can be formed by multiple parallel panels in order to improve the test security and the exposure rate of items and modules. When an individual takes a test, one of the panels is selected,

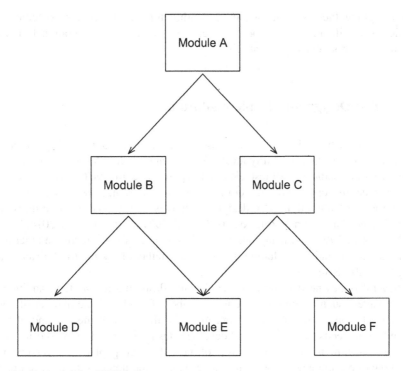

Fig. 6.1 Example of an MST design

and the test taker responds to a subset of the modules defined by a path within this panel (for example, A-C-F in Fig. 6.1.)

In this book, only one panel is considered for the description of the MST. In the operational administration of large-scale standardized tests, multiple panels are deployed to different subgroups of test takers.

In order to design an MST for a specific purpose, several design characteristics need to be investigated and simulated before making a final decision. For example, the number of items in each module, the number of modules at each stage, and the number of stages depend on the desired test length, content, accuracy for ranking or classification and are all subject to research studies conducted to simulate the specific testing situation. This is actually one of the main purposes for having a software package as **mstR** (to be described in the next chapter) that is flexible and supports the analyses prior to and during the implementation of the MST design.

In addition to the design of the MST or each MST panel, one also needs to design the item and/or module and/or panel pool(s) from which specific items are automatically selected for being included into specific modules and panels, or from which pre-assembled modules are directly selected from a pool of modules in order to assembly the panels, or from which pre-assembled panels are selected for specific subgroups of test takers. This hierarchical automatic procedure for adaptation and for designing the test units while accounting for various content specification, difficulty specification, and exposure rate, makes a large-scale operational MST an engineering product (see Luecht, 2014).

6.3   Test Design and Implementation

One design consideration for MST focuses on the accuracy of the ability estimation for a range of test takers' ability levels. Another approach to MST aims to improve accuracy of classification of test takers into appropriate groups. Current operational MSTs generally operate in the item response theory framework; IRT is used as a basis for item calibration, pool design, module assembly, routing algorithms, and scoring (see Hambleton & Zenisky, 2013; Lord, 1971a; Luecht, 1998, 2014; Luecht & Nungester, 1998, and many others). As in the case of CAT, there are other non-IRT based alternatives including a tree-based methodology for MST (Yan et al., 2014, see also Sect. 6.8).

Several issues need to be considered while designing an MST. In addition to basic matters such as length of the test, number of modules, and number of items per module, one also needs to design the difficulty level of the test and of each module, one needs to decide on and develop rules for how to develop scoring rules for the test, how to design the item pool, how the item pool will be replenished, and how to put the new items on the same scale as the items in the pool (a linking strategy for the pool), and eventually, how to link the cut scores across multiple panels if the cut-score is based on an observed-number correct. Other issues include determining mathematical models for item selection and estimation of items and person parameters, for test assembling, for content balancing requirements, and for controlling of item exposures. In order to investigate the optimal design strategies with respect to all these facets, one needs to conduct extensive simulations and evaluate the results according to the goals of the test. Again, all of these can be studied in a straightforward manner with a reliable and easy-to use-software package like **mstR**.

The implementation of an IRT-MST can be separated in two parts: the assembly and the application of the MST. To assemble an MST, sets of items are assembled as modules following some criteria. Then, panels are constructed using those modules. It is also necessary to specify the routing rules.

Module assembly methods of an MST can be made by selecting some set of items that maximizes or surpasses thresholds values of some selected information measurement for fixed ability, θ's. The item set may also need to satisfy all test constraints. The most common information measurement used in MST is the Fisher information. However, other approaches exist, such as Kullback-Leibler information (KL) or the Continuous Entropy Method (CEM), which was implemented in cognitive diagnostic CAT (Cheng, 2009; Xu, Chang, & Douglas, 2003). Other CAT-based approaches of likelihood-weighted module information (LWMI) and posterior-weighted module information (PWMI) also exist in the MST context.

The modules can be assembled by maximizing/minimizing an objective function using linear programming.The Fisher information and the KL criteria for selection were described in Sect. 3.6.3.

6.4   Test Assembly

Test assembly in MST uses statistical and non-statistical requirements, a calibrated item bank and computer programs based on mathematical optimization procedures. The automatic item selection and the optimization of modules and panels are required to meet these constraints. The research on automated test assembly (ATA) (Hambleton & Zenisky, 2013) spans the methods for creating target specifications, positioning of target relative to cut scores on each module, statistical qualities of modules and strategies for integrating constraints. However, since this chapter focuses on only one panel, this step will not be elaborated here. The reader is referred to Yan et al. (2014) for chapters focused on test assembly.

6.5   The Item Bank

In this chapter we do not dwell into the details of the item bank, because the **mstR** package assumes that an item bank already exists. We refer the reader to Veldkamp (2014) for details. Here we briefly outline the process and considerations in designing and populating an item bank.

There are various tasks associated with creating an item bank for a test. The process starts by obtaining the test and module specifications. Once a sufficient number of items in each content category has been created, the test specialists review the item quality. Calibration of the reviewed items takes place after performing the initial pretesting of the newly written items on representative samples of test takers. The item quality is reviewed using statistical and sensitivity analyzes. An ongoing process involves the evaluation and reevaluation of the item bank for size, specifications and content balance. The MST results hinge on the item bank quality relative to measurement goals. The MST design does not overly use the highly discriminating items as may be the case in a CAT. Nevertheless, the MST requires a detailed test blueprint, specific statistical and non-statistical targets, sufficient items in the pool and sufficient sample sizes of test takers for the item calibration.

Matching an item bank to test design choices is among the key processes in assessment. This overarching goal presents several opportunities for continuing research such as on item and test security strategies, item bank composition, automated test assembly, and integration of item format and the inclusion of automated scoring.

6.6   IRT-Based MST

This section provides a brief overview of the most important aspects of IRT-based MST.

6.6.1   Module Selection

The selection of the next module is based on comparing the current estimated ability (or test) score to predefined thresholds (or cut-scores) and then, on the rules for selecting the next module according to such thresholds. However, in an IRT approach to MST, it is better to select the next module by using an information related criterion, for instance based on Fisher information. Most of these criteria can be derived from the CAT context (see Sect. 3.6.3) and make use of Fisher or Kullback-Leibler information computed on the whole set of items within the module. Extensions of the likelihood weighted information and posterior-weighted information functions used in CAT can also be considered in MST. The Continuous Entropy method (CEM), which was implemented in cognitive diagnostic CAT (Xu et al., 2003), can also be used.

For the Rasch and 2PL models, items have the most information when the difficulty parameter and the latent trait have the same value, which is also where the probability of a correct response is equal to, 50%.

6.6.2   Routing

The routing rules can be implemented using several approaches, including the cut-scores based on the number of correct scores or according to cut-off points for the θ estimates. These cut-off points can be based on information functions or according to the latent trait distribution.

The routing criterion can be expressed in terms of the ability estimate, and if an EAP is used, the criterion can be designed to be independent of a specific sample at a specific test administration. This can be convenient for an operational MST, in which cohort effects may exist.

This cut-score may be converted to an administration-specific summed score, using the EAP estimate of ability obtained at the end of a stage (Haberman & von Davier, 2014). Furthermore, the criteria that are dependent on the administration (and are not EAP or MAP based) need to be equated across administrators.

6.6.3 Latent Trait Estimation

Multistage testing relies on the idea that the error associated with an ability estimate or a classification decision is a function of the information available for measurement. Tests may be tailored to test takers if information contributions of test items can be determined as functions of test taker's ability. IRT provides methods for determining information contributions. Another asset of MST is that an item score in the operational section is independent of any other item score from a stage before and at the stage at which the item was presented given the test taker's ability, despite the routing (see Eggen & Verhelst, 2011; Mislevy & Chang, 2000).

As for CAT purposes, various ability estimators are available: maximum likelihood, Bayes modal (or maximum a posteriori), expected a posteriori and weighted likelihood are the most common ones. These methods were described in detail in Sect. 2.3.2. In MST, such estimates are computed at the end of the whole module administration. Moreover, provisional ability estimates within the module are not necessary anymore, so the estimation step is faster than in CAT. Moreover, if the routing module contains items with sufficient spread of difficulty across the ability scale, there is less chance of observing a constant pattern (i.e., all correct or all incorrect responses to dichotomous items), hence adjustment methods for constant patterns (as described in e.g., Sect. 4.5.3) are less likely to be required than in CAT scenarios.

As presented in Sect. 2.1, one of the main assumptions of traditional IRT models is local item independence (LI). The LI property ensures that the estimation of the IRT model parameters is not impacted by the routing of test takers to a stage given their ability at the previous stage. The consequences of routing on LI for adaptive tests such as MST are not substantial (Eggen & Verhelst, 2011; Mislevy & Chang, 2000) for EAP or MAP estimators when using a generalized-partial credit IRT model. For a ML estimator combined with the use of the 3PL IRT model, the consequences of routing are more problematic (Haberman & von Davier, 2014). The MLE relies on the central limit theorem for martingales to cope with the routing rules, it requires a large number of items, and is not defined for those test takers with extreme scores (von Davier & Haberman, 2014).

Asymptotically (Holland, 1990), the MAP, EAP, and MLE for stage j are close to each other if the initial j stages always have a large number of items. This property supports the recommendation of longer modules for an MST (if the estimator is not EAP), which leads to a dilemma for test developers, as one of the goals of the MST is to allow for a shorter testing time.

In addition to these methodologies, Haberman and von Davier (2014) investigated the relationship between the IRT estimates and the total number correct or summed scores and showed that the estimators based on the number correct scores have the desired asymptotic properties. One other way to score an MST is to use a number correct score and a tree-based regression or classification for routing, as described later in this chapter.

6.7   IRT Linking

Linking of item parameters is necessary in operational programs, in which different items are used at each test administration and new items need to be included in the item pool. The test administration and the test specifications are standardized to ensure test comparability over administrations. In order to always have a sufficient number of new items in the item bank, one needs to administer the new items on a representative sample from the test taker population.

Linking in an MST operational program takes place in three phases. In the initial phase of data collection, (conventional) test administrations are used to build the modules and routing rules. The second phase begins when MST administrations start and the data are collected to establish scoring rules, develop new test modules, equate the cut-scores for routing, and ensure comparability of tests over time. In the third phase, when many MST administrations are involved, special procedures are considered for maintenance of a stable linkage and test validity.

Population parameter linking in an operational program is particularly challenging due to the large number of items which may number in thousands. In most testing programs, the method used is the marginal maximum likelihood (MML) applied to observed item responses as if all the items were presented to each test taker.

There are three major approaches to conducting the item calibration and linking: concurrent calibration, sequential linking, and simultaneous linking. Concurrent calibration is usually employed in the start-up period and it is also the basis of pre-equating. Concurrent calibration of all the items is the most efficient method, if the number of items per administration is not very large and if sample sizes are large enough. In the case of accurate normal approximations, the estimates are asymptotically efficient.

However, in an operational program, the number of items and the number of administrations tend to be quite large, which leads to difficulties in estimating the item parameters concurrently. Hence, another approach is used more often, that is sequential linking of separately calibrated item parameters, which is usually employed at each administration (or a batch of administrations) to place the new items on the same scale as the items in the item bank (Stocking & Lord, 1983).

Lastly, simultaneous linking is conducted to adjust separately calibrated parameters and to correct for the measurement and sampling error from the individual calibrations. It is used at regular intervals in the life of an assessment. This method is used only for operational programs that have an almost continuous administration mode, and where the risk of cumulative errors is high (Haberman & von Davier, 2014).

6.8   MST with Regression Trees

Current CATs and MSTs programs rely heavily on IRT. However, CAT does not perform well when the samples are small or IRT assumptions are violated, e.g., unidimensionality (Yan, Lewis, & Stocking, 2004). As an alternative approach, the tree-based CAT algorithm (Yan et al., 2004) seems to perform as well as an IRT-based CAT or MST.

Similarly to the CAT context, a tree-based approach for MST was developed (Yan, Lewis, & von Davier, 2014b). In the Tree-based MST algorithm, module scores refer to the number of correct responses to the items or the summed scores in the module. The criterion score is the total number correct score for a test consisting of all items. The cut-score of the module splits the current sample of test takers into two subsamples to be administered to the easier and more difficult modules at the next stage.

The Tree-based MST is a prediction system that routes test takers efficiently to the appropriate groups, based on their module scores at each stage of testing. It predicts their total scores based the paths they take and the modules they answer, without introducing latent traits or true scores.

The Tree-based MST has three steps (for the three-stage MST considered here): At the first step, one computes the total observed number-correct score for the routing module (Module A) for all test takers. Next for each possible number-correct scores that splits the current sample into two subsamples (nodes) one computes the sum of the within subsample sums of squares of criterion and finds the optimal number-correct cut score for which the within group sum of squares is at the minimum. At the second step, one computes the total observed number-correct score for the modules from Stage 1 and Stage 2. Then for each possible pair of number-correct score that may split the Stage 2 samples into two more subsamples one combines the higher part of lower group and lower part of the higher group and repeats the process from step 1, for the three groups. Hence, one needs to find the pair of optimal number-correct cut scores for which the within group sum squares for the three groups is the smallest. In the third step, one computes multiple linear regressions of criterion on observed number-correct scores for all module scores for each subsample corresponding to one of the four paths.

Though being of primary interest for operational MSTs where underlying assumptions of IRT are violated, this method is not yet implemented in **mstR**. We refer to Yan et al. (2004) and Yan et al. (2014b) for further details.

6.9   Final Comments

This chapter provided an overview of the MST. An MST is a complex test design in which engineering and psychometric features come together. There are numerous practical issues that need to be taken into consideration when developing an MST, and which cannot be explicated in advance. A significant amount of simulations and adjustments are needed to optimize the MST. Below is a list of elements that need numerous repetitive simulations:

- Test design and assembly
- Item pool/bank and maintenance, multiple pools
- Item selection; adding, removing and revising items; out-of-level items
- Maintaining scale consistency
- The response models
- Quality of calibration
- Multidimensionality, Differential Item Functioning, Speededness
- Scoring procedures
- Test security and item exposure

Some of these characteristics are illustrated through simulation examples in Chap. 8.

Chapter 7
Simulations of Computerized Multistage Tests

In the first part of this book, focus was put on computerized adaptive testing and the R package **catR** was presented as a tool to simulate CAT patterns and scenarios. In the context of multistage testing, the R package **mstR** was developed in a similar spirit. This chapter provides a brief description of **mstR**, with emphasis on main differences between **mstR** and **catR** for specific MST purposes.

7.1   The R Package mstR

Within the framework of CAT, the R package **catR** was probably the first package devoted to adaptive testing and was made publicly available through CRAN (its first version was released in June 2010). Since then, other CAT-related packages were developed and published on CRAN, for instance **catIrt** (Nydick, 2014), available in August 2012, and **mirtCAT** (Chalmers, 2016), available in August 2014. However, in the context of multistage testing, the R package **mstR** is, to the best of our knowledge, currently the only one available from CRAN. Moreover, it was built with the same general structure as **catR**, which brings two assets. First, there are many common functions in both packages, which greatly simplifies the learning and manipulations of these routines. Second, both packages share the same logic and require the same input files and arguments. Of course, **catR** and **mstR** hold their own specific functions and options. Those of **mstR** are described in this chapter.

The general architecture of **mstR** looks pretty similar to that of **catR** and is schematically displayed in Fig. 7.1. The similarities of Figs. 7.1 and 4.1 are straightforward: they contain the same basic IRT functions and IRT scoring functions, together with specific functions for MST (such as next module selection) and one top-level generating function randomMST(). The testListMST() function was also created to internally check the consistency of the various input lists to randomMST() calls.

© Springer International Publishing AG 2017

D. Magis et al., *Computerized Adaptive and Multistage Testing with R*, Use R!,

https://doi.org/10.1007/978-3-319-69218-0_7

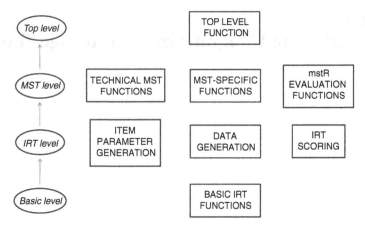

Fig. 7.1 General structure of the mstR package

The basic IRT functions Pi(), Ii() and Ji(), the item parameter generation functions genDichoMatrix() and genPolyMatrix(), the data generation function dataGen(), and the IRT scoring functions thetaEst() and semTheta() are described respectively in Sects. 4.4, 4.5.1, 4.5.2, and 4.5.3. Please refer to these sections for detailed descriptions, as they work similarly in the package **mstR**. Therefore, we focus on specific MST elements, starting with the proper design for modules and stages.

7.2 Multistage Structure for Item Banks

In a CAT design, the item bank is a matrix of item parameters calibrated under an appropriate IRT model. This item bank can have some structure for content balancing by adding a variable for sub-group membership (see Sect. 4.6.2). In a MST design, however, the item bank is actually a set of predefined *modules*, with each module belonging to one and only one *stage* of the design (though some items may belong to more than one module and hence to different stages). A path which allows to pass from one module in one stage to another module in the next stage, must also be provided.

In **mstR**, items are still provided in a matrix of suitable item parameters, similarly as in the **catR** package, and the modules are defined with an appropriate item-by-module *design matrix*. This matrix has as many rows as the number of items and one column per module. Matrix entries are zeros and ones, a value of one on row i and column j indicates that item i belongs to module j. As mentioned above, one item can belong to more than one module so multiple "1" entries per row are allowed.

To illustrate this module definition, let us consider the following artificial example (to be considered throughout this chapter). A multistage test of the format

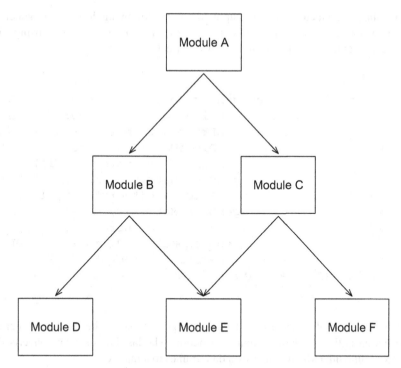

Fig. 7.2 An example of an 1-2-3 MST design

1-2-3 will be simulated. For simplicity, the module from stage 1 is referred to as module A, modules from stage 2 as modules B and C, and modules from stage 3 as modules D, E and F. Moreover, paths will be allowed as follows: from module A to both modules B and C (that is all possible paths from stage 1 to stage 2); from module B to modules D and E; and from module C to modules E and F. Figure 7.2 depicts this artificial design.

Now, module A consists of 8 items, modules B and C of 6 items each, and modules D–F of 9 items each, leading to an item bank of 47 items. For the sake of simplicity, the first 8 items of the bank will be those from module A, the next 6 items from module B, and so on. Finally, all items are generated under the dichotomous 2PL model (2.5), using the default parent distribution for item discrimination levels (i.e., normal distribution with mean 1 and standard deviation 0.2), while item difficulties are drawn from the normal distribution with standard deviation 1 but various mean values according to the module in question: $N(0, 1)$ for module A; $N(-1, 1)$ and $N(1, 1)$ for modules B and C respectively; and $N(-2, 1)$, $N(0, 1)$ and $N(2, 1)$ for modules D, E and F, respectively. In this way, one specifies the following: (a) module A has average difficulty items in stage 1; (b) module B holds easier items than module C in stage 2; (c) modules D, E and F contain very easy, average, and very difficult items, respectively.

All these parameters can be supplied to the following R code to generate the item bank, which will be stored in the R object it.MST, using the genDichoMatrix() function described in Sect. 4.5.1:

```
R> it.MST <- rbind(genDichoMatrix(8, model = "2PL"),
+                  genDichoMatrix(6, model = "2PL",
+                    bPrior = c("norm", -1, 1)),
+                  genDichoMatrix(6, model = "2PL",
+                    bPrior = c("norm", 1, 1)),
+                  genDichoMatrix(9, model = "2PL",
+                    bPrior = c("norm", -2, 1)),
+                  genDichoMatrix(9, model = "2PL",
+                    bPrior = c("norm", 0, 1)),
+                  genDichoMatrix(9, model = "2PL",
+                    bPrior = c("norm", 2, 1)))
R> it.MST <- as.matrix(it.MST)
```

The first line of code has no bPrior argument since the by-default parent distribution is the standard normal distribution. The last line of code converts the created item bank (which comes as a data frame) to a matrix.

Now, to specify the different modules, an appropriate design matrix with 47 rows and 6 columns (called modules) must be created. The first column will correspond to module A, the second column to module B, etc. and each column must be filled with "1" values for adequate items in the it.MST bank. This yields the following code:

```
R> modules <- matrix(0, 47, 6)
R> modules[1:8, 1] <- 1
R> modules[9:14, 2] <- 1
R> modules[15:20, 3] <- 1
R> modules[21:29, 4] <- 1
R> modules[30:38, 5] <- 1
R> modules[39:47, 6] <- 1
```

In other words, the first 8 items belong to module A, the next 6 items to module B etc., just as expected. A quick check of the number of items by module can be obtained as follows:

```
R> colSums(modules)
[1] 8 6 6 9 9 9
```

Furthermore, rowSums(modules) will return only "1" values, indicating that in this design, each item belongs to only one module.

The final step towards defining the appropriate MST structure is to specify which paths between modules will be allowed by the algorithm. In **mstR**, this is solved by specifying a suitable $R \times R$ *transition matrix* (with R being the number of modules), that is, a square matrix of binary entries, in which a value "1" on row i and column j indicates that there exists a path from module i to module j. It is important that the numbering of the modules in this transition matrix (i.e., the row/column numbering) be identical to the numbering of the columns in the design matrix for setting up the modules. In other words, the first row/column of the transition matrix corresponds to the module set in the first column of the modules' design matrix, and so on.

In this example, the following R code is required to create the transition matrix trans:

```
R> trans <- matrix(0, 6, 6)
R> trans[1, 2:3] <- 1
R> trans[2, 4:5] <- 1
R> trans[3, 5:6] <- 1
```

which has the following format:

```
R> trans
      [,1] [,2] [,3] [,4] [,5] [,6]
[1,]    0    1    1    0    0    0
[2,]    0    0    0    1    1    0
[3,]    0    0    0    0    1    1
[4,]    0    0    0    0    0    0
[5,]    0    0    0    0    0    0
[6,]    0    0    0    0    0    0
```

The first column has only zeros, meaning that there is no path pointing to module A, since it is the module from the first stage. Similarly, there are only zeros in the last three rows of the trans matrix, indicating that there is no path departing from modules D, E and F (as expected since they compose the last stage). The two "1" values in row 1 and columns 2 and 3 indicate that there are two paths departing from module A, pointing to modules B and C, respectively. Other "1" values can be interpreted similarly.

These three elements (the item bank it.MST, the design matrix modules and the transition matrix trans) are the basic ingredients to the most important MST functions in **mstR**. These elements are part of the input arguments in **mstR** in exactly the same way as in **catR** functions.

7.3 MST Functions

The two MST-specific functions in **mstR** are startModule() to select the module from the first stage, and nextModule() to select the next module in the next stage, according to currently recorded item responses. They perform in the same way as functions startItems() and nextItem() in **catR** and they also share the same logic.

7.3.1 The startModule() Function

First, the startModule() function takes four basic input arguments:

- itemBank to set the matrix of item parameters;
- modules to specify the design matrix of modules;
- transMatrix to provide the transition matrix between all modules;
- model to mention which IRT model was considered for item calibration.

The latter works the same way as in the **catR** package, that is, it takes value NULL for dichotomous IRT models and the appropriate acronym for the polytomous IRT models (see Sect. 4.2 for the list of available models and acronyms). Moreover, startModule() has three options to select the first module to be administered (among all available modules from the first stage): (a) user-specified; (b) random selection; (c) maximizing the module information. These options are set with the arguments fixModule and seed and as follows:

1. If fixModule is set to NULL (which is the by-default value) then the first module is not chosen by the user or administrator. Otherwise fixModule must be supplied with an integer value that specifies the module to be chosen. Note that the module number identification corresponds to the column number of this module in the modules design matrix.
2. If fixModule is set to NULL and seed is not NULL, then random selection of the first module is performed. The random selection can be controlled via a user-supplied seed number to be provided to the seed argument, while completely random selection (without fixing the random seed) is done by setting seed to NA. By default, seed is NULL so no random selection is performed.

3. Eventually, if both fixModule and seed are set to NULL, then the first module is chosen as the most informative module with respect to a predefined ability level that is provided to the theta argument. Module information is defined as the sum of all individual information functions of the items belonging to the module. By default the theta argument has value zero.

The output of startModule() function is a list which consists of the chosen module (specified by its identification number), the items belonging to this module (also identified by their number in the item bank), the corresponding item parameters for the module, and the starting ability level (or NA if the module is chosen without maximizing the information function).

To illustrate this function, let us select the first module from our illustrative MST design. Since there is only one module available in the first stage (i.e., module A), only this module can be selected, regardless of the seed value chosen for random selection or the initial ability level. Hence, by keeping all the default arguments for the startModule() function,

```
R> startModule(itemBank = it.MST, modules = modules,
              transMatrix = trans)
```

one can obtain the following output:

```
$module
[1] 1

$items
[1] 1 2 3 4 5 6 7 8

$par
            a           b c d
1 1.1151563 -0.6264538 0 1
2 0.9389223  0.1836433 0 1
3 1.3023562 -0.8356286 0 1
4 1.0779686  1.5952808 0 1
5 0.8757519  0.3295078 0 1
6 0.5570600 -0.8204684 0 1
7 1.2249862  0.4874291 0 1
8 0.9910133  0.7383247 0 1
```

(continued)

```
$thStart
[1]  0
```

Adding for instance seed = 1 to the code will return exactly the same output
(for the reason stated above), except that the $thStart argument now takes the
value NA. Eventually, if one misspecifies the module number and tries to select one
module from another stage (say module C, which is the third recorded module in
the design matrix), then the following code,

```
R> startModule(itemBank = it.MST, modules = modules,
              transMatrix = trans, fixModule = 3)
```

will return the following warning message:

```
Error: Selected module is not from stage 1!
```

7.3.2 The nextModule() Function

The second MST-specific function is nextModule() and is used to switch from
one stage to the next one by selecting the most appropriate module from the set
of available ones at the next stage. As for the startModule() function, it
requires the input arguments itemBank, modules, transMatrix, and model
for appropriate setting. Moreover, the two input arguments current.module
and out must be supplied, the former with the identification number of the last
administered module (this is to allow proper selection of the module in the next
stage) and the latter as a vector of previously administered items (in all previously
selected modules). Since items may belong to more than one module, the argument
out is important and can be used to avoid selecting a module with previously
administered items. Finally, one also has to specify the current ability estimate (or
score) for the next module selection. This is done by using the argument theta.

There are two main approaches to select the next module: an appropriate optimal
selection criterion, or total test scores and suitable cut-off values between the
modules. In the former approach, six selection criteria are currently available
and can be specified with the criterion argument. Five criteria constitute a
straightforward extension of their CAT versions into the MST context:

- *maximum Fisher module information* (or "MFI"), by maximizing the module information function;
- *maximum likelihood weighted module information* (or "MLWMI"), by maximizing the module information function weighted by the likelihood function (of currently administered items);
- *maximum posterior weighted module information* (or "MPWMI"), by maximizing the module information function weighted by the posterior distribution;
- *maximum module Kullback-Leibler information* (or "MKL"), by maximizing the module Kullback-Leibler information function (weighted by the likelihood function);
- *maximum posterior module Kullback-Leibler information* (or "MKLP"), by maximizing the module Kullback-Leibler information function (weighted by the posterior distribution).

The sixth option is the *random selection*, set by the "random" value for the criterion argument.

Additional tuning arguments for the various methods can be supplied through the arguments priorDist, priorPar, D, range, and parInt. They work identically (and take the same allowed values) as the corresponding arguments in **catR** package, please refer to Sect. 4.6.2 for further details. The output of nextModule() is a list with, among others, the identification number of the chosen module ($module), the items that belong to the chosen module ($items) and the corresponding item parameters ($par).

Let us illustrate this selection process by using the artificial item bank and MST design. The module from the first stage is module A with 8 items. For the sake of simplicity, let us assume that the test taker answered the first four items correctly and the last four items incorrectly. The provisional ability estimate, computed by maximum likelihood, is obtained with the following code (and the thetaEst() function):

```
R> x <- c(1, 1, 1, 1, 0, 0, 0, 0)
R> thetaEst(it.MST[1:8, ], x, method = "ML")
[1] 0.3654008
```

Now, the module from stage 2 will be chosen as the most informative module for the current ability estimate (i.e.,) as follows:

```
R> nextModule(itemBank = it.MST, modules = modules,
+             transMatrix = trans, current.module = 1,
```

(continued)

```
+                    out = 1:8, criterion = "MFI",
+                    x = x, theta = th)

$module
[1] 3

$items
[1] 15 16 17 18 19 20

$par
            a           b c d
15 1.0974858 0.3735462 0 1
16 1.1476649 1.1836433 0 1
17 1.1151563 0.1643714 0 1
18 0.9389223 2.5952808 0 1
19 1.3023562 1.3295078 0 1
20 1.0779686 0.1795316 0 1
```

From this output, it appears that module 3 (i.e., module C) is chosen as the most informative module (among the two available modules in stage 2) for the current ability estimate. A quick check with all other methods (but the random one) yield the same decision for this example.

These criterion-based module selection methods rely on the IRT information functions and some current ability level estimates. However, unlike CAT framework, it is also possible to set up module selection methods based on predefined *cut-off values* and to choose the next module as the one whose provisional ability estimate lies between the two corresponding cut-off points. The merit of this method is that it works both for traditional IRT estimates and for total test scores, that is, the sum of scores of currently administered items. The drawback is that these cut-off values must be supplied a priori by the test developer.

To make use of this approach, the cutoff argument must be set to an appropriate threshold matrix with as many rows as the total number of modules in the bank, and two columns. Each row contains the lower and upper thresholds for the corresponding score: if, for some module m, the corresponding thresholds are t_{m1} and t_{m2}, then module m will be chosen for the next stage if the current ability estimate (say $\hat{\theta}$) is such that $t_{m1} \leq \hat{\theta} < t_{m2}$. In practice, infinite values (i.e., $t_{m1} = -\infty$ and $t_{m2} = +\infty$) are allowed, and for compatibility the thresholds must be supplied for all modules in the bank, including those from the first stage that will not be selected using nextModule(). It is important to note that once the cutoff matrix is supplied to the nextModule() function, then the criterion argument becomes irrelevant and the module selection will be based on the predefined thresholds only.

To illustrate this scenario, consider the artificial MST design with six modules and the following cut-off matrix cut:

```
R> cut <- matrix(NA, 6, 2)
R> cut[2,] <- c(-Inf, 0)
R> cut[3,] <- c(0, Inf)
R> cut[4,] <- c(-Inf, -1)
R> cut[5,] <- c(-1, 1)
R> cut[6,] <- c(1, Inf)
R> cut
       [,1] [,2]
[1,]    NA    NA
[2,]  -Inf     0
[3,]     0   Inf
[4,]  -Inf    -1
[5,]    -1     1
[6,]     1   Inf
```

This matrix can be interpreted as follows. The first row holds NA values but it corresponds to module A from the first stage, so any value in this first row is allowed (the thresholds must be supplied for compatibility but will not be used anymore in this function). Modules B and C (from stage 2) have respective sets of thresholds $(-\infty, 0)$ and $[0, +\infty)$, which means that if the provisional ability estimate $\hat{\theta}$ is strictly smaller than zero, then module B will be chosen, otherwise module C will be chosen. Finally, the thresholds for modules D, E and F (from stage 3) set the cut-off values to -1 and 1, so that module D is chosen if $\hat{\theta} < -1$, module E is chosen if $-1 \leq \hat{\theta} < 1$ and module F is chosen if $\hat{\theta} \geq 1$. Though not mandatory, it is strongly advised that values -Inf and Inf are supplied to denote that the module is chosen for very small (or very large) ability estimates.

To illustrate the use of this cut matrix, we re-consider the former example by selecting now the next module on the basis of the current ability estimate and those fixed thresholds:

```
R> nextModule(itemBank = it.MST, modules = modules,
+             transMatrix = trans, current.module = 1,
+             out = 1:8, cutoff = cut, theta = th)

$module
[1] 3
```

(continued)

```
$items
[1] 15 16 17 18 19 20

$par
         a         b c d
15 1.0974858 0.3735462 0 1
16 1.1476649 1.1836433 0 1
17 1.1151563 0.1643714 0 1
18 0.9389223 2.5952808 0 1
19 1.3023562 1.3295078 0 1
20 1.0779686 0.1795316 0 1
```

Here also module C is chosen, as expected (the provisional ability estimate is 0.365 and is larger than the threshold of zero). If we assume (for the sake of illustration) that the current ability estimate is −0.2 instead, then module B will be chosen, as illustrated below:

```
R> nextModule(itemBank = it.MST, modules = modules,
+            transMatrix = trans, current.module = 1,
+            out = 1:8, cutoff = cut, theta = -0.2)

$module
[1] 2

$items
[1]  9 10 11 12 13 14

$par
         a         b c d
9  1.0974858 -1.6264538 0 1
10 1.1476649 -0.8163567 0 1
11 1.1151563 -1.8356286 0 1
12 0.9389223  0.5952808 0 1
13 1.3023562 -0.6704922 0 1
14 1.0779686 -1.8204684 0 1
```

The advantage of this approach is that one can work with total test scores instead of IRT ability estimates, and provide integer thresholds to distinguish between the different modules from the same stage (though numeric values are also allowed, making use of integer thresholds provides a more comprehensive choice of cut-off values in this context). To illustrate this approach, we create the cut.score matrix to represent the following thresholds:

1. at stage 2, the threshold is set to 5, so that module B is chosen if the provisional test score (based on module A only) is (strictly) smaller than 5 (that is, from 0 to 4) and module C otherwise;
2. at stage 3, the thresholds between modules D, E, and F are set to 5 and 10, so that modules D, E, and F are chosen if the provisional test score (based on the two modules from previous stages) falls between 0 and 4, between 5 and 9, and between 10 and 14, respectively.

This scenario is implemented with the following code:

```
R> cut.score <- matrix(NA, 6, 2)
R> cut.score [2,] <- c(0, 5)
R> cut.score [3,] <- c(5, 8)
R> cut.score [4,] <- c(0, 5)
R> cut.score [5,] <- c(5, 10)
R> cut.score [6,] <- c(10, 14)
R> cut.score
      [,1] [,2]
[1,]   NA   NA
[2,]    0    5
[3,]    5    8
[4,]    0    5
[5,]    5   10
[6,]   10   14
```

Note that the lower cut-off values for modules B and D could be fixed to -Inf, as well as the upper cut-off values for modules C and F to Inf, without impacting the process. However, it may be easier to interpret such integer cut-off values in the case of test scores (instead of IRT ability estimates).

As a final illustration, let us select the module in the third stage, using the cut.score matrix provided above and with provisional test scores of 11 from modules A and C. The current module is therefore module C, referred to as identification number 3. Recall that module A contains items 1–8 and module C contains items 15–20 (as numbered in the item bank). The following code yields the expected answer (i.e., module F is chosen):

```
R> nextModule(itemBank = it.MST, modules = modules,
+              transMatrix = trans, current.module = 3,
```

(continued)

```
+                    out = c(1:8, 15:20), cutoff = cut.score,
+                    theta = 11)
$module
[1]  6


$items
[1]  39 40 41 42 43 44 45 46 47


$par
              a          b c d
39 0.9389223 1.373546 0 1
40 1.3023562 2.183643 0 1
41 1.0779686 1.164371 0 1
42 0.8757519 3.595281 0 1
43 0.5570600 2.329508 0 1
44 1.2249862 1.179532 0 1
45 0.9910133 2.487429 0 1
46 0.9967619 2.738325 0 1
47 1.1887672 2.575781 0 1
```

7.4 The `randomMST()` Function

The main function in the **mstR** package is the `randomMST()` function, which is similar to function `randomCAT()` in package **catR**. Both are conceptually identical in terms of their input arguments: lists of options to set the various steps of the adaptive test, specific input arguments for specifying the item bank, and options for saving the output. The main differences are that `randomMST()` requires more input arguments in order to correctly specify the structure of the MST, but does not require a `stop` list as the test stops once the module from the final stage is administered and item responses are recorded.

This section focuses on the input values in detail, similar to the approach in Sect. 4.7 for the `randomCAT()` function.

7.4.1 Input Information

Table 7.1 lists all available input arguments of the `randomMST()` function.

Most of the input arguments (`model`, `responses`, `genSeed`, `allTheta`, `save.output` and `output`) are common to the `randomCAT()` function and

Table 7.1 Input arguments for randomMST() function

Name	Function	Type
trueTheta	Sets the true ability level	Numeric
itemBank	Sets the matrix of item parameters	Matrix
modules	Sets the MST design matrix	Binary matrix
transMatrix	Sets the transition matrix	Binary matrix
model	Sets the type of polytomous IRT model	Model acronym or NULL
responses	Provides item responses for post-hoc simulations	Vector or NULL
genSeed	Fixes the general random seed	Numeric or NULL
start	Sets the options to select the first module	Appropriate list
test	Sets the options for provisional ability estimation and next module selection	Appropriate list
final	Sets the options for final ability estimation	Appropriate list
allTheta	Sets whether all ability estimates must be returned	Logical
save.output	Sets whether output should be saved	Logical
output	Sets the output location, file name and extension	Appropriate vector

Table 7.2 Elements of the start list of function randomMST()

Name	Function	Type	Default
fixModule	Fixes the module to administer	Integer or NULL	NULL
seed	Sets the random seed to sample the starting module	Numeric or NA or NULL	NULL
theta	Sets the starting ability level	Numeric value	0
D	Sets the scaling constant D	Numeric	1

take the same values. The start, test, and final lists perform similarly but with possibly different elements; they are further described later on. The modules and transMatrix arguments are mandatory to define the MST design and to allow transitions between modules. Finally, some arguments of randomCAT(), such as nAvailable and cbControl, are not available in randomMST().

7.4.2 The start List

The start list has four possible elements that are listed and described in Table 7.2.

Scaling constant D is used only if the first module is selected as the most informative one among available modules from the first stage. Moreover, the method for choosing the first module is determined in the following sequential order.

1. If fixModule takes an integer value (i.e., the identification number of the module) then this module will be chosen and elements seed and theta are ignored.

Table 7.3 Elements of the `test` list of function `randomMST()`

Name	Function	Type	Default
method	Sets the provisional ability estimator	Appropriate acronym	"BM"
priorDist	Sets the prior distribution for ability estimation	Appropriate acronym	"norm"
priorPar	Sets the parameters of the prior distribution	Numeric vector	c(0,1)
range	Sets the range of ability estimation	Numeric vector	c(-4,4)
D	Sets the scaling constant D	Numeric	1
parInt	Sets the sequence of quadrature points	Numeric vector	c(-4,4,33)
moduleSelect	Sets the method for next module selection	Appropriate acronym	"MFI"
constantPatt	Sets the method to deal with constant patterns	Appropriate acronym or NULL	NULL
cutoff	Sets an appropriate matrix of cut-off values	Appropriate matrix or NULL	NULL

2. If `fixModule` is `NULL` and `seed` is either a numeric value or `NA`, then random selection of the first module is performed, and the random draw is either fixed by the numeric `seed` value or completely random (when value `NA` is provided). Regardless of the value of `seed`, element `theta` is ignored in this case.
3. If both `fixModule` and `seed` are set to `NULL`, then the most informative module is chosen according to the starting ability level fixed by `theta` (zero by default).

7.4.3   The test List

The `test` list also contains many common elements of the corresponding list for `randomCAT()` function (listed in Table 4.5). For the present case, various allowed elements are listed in Table 7.3.

First, elements `method`, `priorDist`, `priorPar`, `range`, `D`, `parInt`, and `constantPatt` work identically and take exactly the same possible values as the corresponding elements in the `test` list of the `randomCAT()` function (see Sect. 4.7 for further details). There is, however, one notable exception: in **mstR** the `method` argument also allows the `"score"` value, in order to compute test scores instead of IRT ability estimates. This `"score"` value should be considered when specifying cut-off matrix of test scores (see Sect. 7.3.2).

Second, element `itemSelect` is replaced by `moduleSelect` and takes possible values `"MFI"`, `"MLWMI"`, `"MPWMI"`, `"MKL"`, `"MKLP"`, and `"random"`, as described in Sect. 7.3 for the `nextModule()` function. Finally, the MST-specific cutoff element is present to provide (ability-based or score-based) thresholds for the modules from each stage. The appropriate format for the cutoff matrix is described in Sect. 7.3 and, whenever provided, it disables the element moduleSelect.

7.4.4 The final List

The final list has the same applications and allowed values as in the test list (listed in Table 7.3): method, priorDist, priorPar, range, D and parInt. In addition, element alpha defines the significance level that is used to compute the final confidence interval for the final ability estimation. By default it is set to be 0.05 but can be modified to any numeric value between zero and one.

7.4.5 Output Information

The output of randomMST() is a list of class "mst" with many elements. Among others, many input arguments are included in the output list, together with the set of selected modules ($selected.module), the whole set of administered items ($testItems) and the corresponding parameters ($itemPar), the full response pattern ($pattern), the set of provisional ability estimates or test scores ($thetaProv) and corresponding standard errors ($seProv) in case of IRT scoring rules, the final ability estimate or score ($thFinal) and its related standard error in case of IRT scoring ($seFinal). We refer interested readers to the help file of the randomMST() function for a complete list of output elements.

This output can be displayed in a more readable way by using the print.mst() S3 function. It displays first a summary of the chosen MST options and the MST structure (determined by the input modules and transMatrix arguments), followed by a summary of each module administration at its stage. These summaries are presented in successive tables with as many columns as the number of items in the module, and five rows: the item number in the module (from one to the module's length), the item identification number (in the item bank), the item response, the provisional ability estimate, and its related standard error. Note that the latter two rows are displayed only if IRT scoring was required (and not total test scores) and if the argument allTheta was set to TRUE in the randomMST() function. Finally, summary results at the end of the test (final ability or score estimate and related information) are provided. This output can be saved as an external file by specifying the arguments save.output and output as explained in Sect. 7.4.1.

Finally, the full MST design and the related path derived from the use of randomMST() can be graphically displayed using the S3 function plot.mst(). One input argument of plot.mst() is the output of randomMST() and other optional arguments are listed in Table 7.4.

Table 7.4 Elements of the `plot.mst()` function

Name	Usage	Type	Default
x	Sets input list for plotting	Object of class "mst"	NA
show.path	Highlight selected path in plot?	Logical	TRUE
border.col	Specifies the color of module borders	Color name	"red"
arrow.col	Specifies the color of arrows of selected path	Color name	"red"
module.names	Sets the module names to display	Vector of characters or NULL	NULL
save.plot	Should the plot be saved?	Logical	FALSE
save.options	Location and format of saved plot	Vector of appropriate character strings	c("path", "name", "pdf")

By default the resulting plot will look like Fig. 7.2, that is, all modules from the same stage are displayed in a row, the first stage on the top of the plot and the last stage at the bottom of the plot. In addition, the selected modules during the MST generation are highlighted by a colored box (default color is red and can be modified with `border.col` argument), and the arrows connecting the selected modules are also highlighted in color (with default value red also, and can be modified using `arrow.col` argument). It is however possible to remove the highlighting of the selected path across modules and stages by setting argument `show.path` to `FALSE`. Finally, by default the modules are labeled with their identification number in the `modules` matrix, but character strings can be displayed instead by specifying an appropriate vector of characters to the argument `module.names`. The remaining arguments `save.plot` and `save.options` are similar to those of the `plot.cat()` function and were described in Sect. 4.7.6.

Chapter 8
Examples of Simulations Using *mstR*

In this chapter, we present several examples of using the R package **mstR** which is similar to CAT but specific for MST. We then implement a study with several MST designs, and an empirical comparison of MST versus CAT.

8.1 Introduction

For the purpose of illustrating the package **mstR**, the 2PL item bank described in Sect. 5.1.1 will be considered once again. In addition, several appropriate MST structures will be used in the simulations.

In all structures, the MST design is always set as a 1-2-3 design (i.e., three stages of respectively one, two, and three modules). The composition of the modules varies from one design to another and were originally used by Yan, Lewis, and von Davier (2014a, 2014b). Beyond the detailed composition of each module in terms of items in the 2PL item bank, the size of the modules at each stage is varied with the constraint that the total test length must be equal to 45. Table 8.1 lists the respective sizes of each module in each stage, for the six available MST designs to be compared.

Table 8.1 indicates that the general designs 1–3 are identical to designs 4–6 respectively, in terms of module lengths per stage. Only the compositions and structures of the modules differ. Designs 1 and 4 have equal module lengths on the whole MST. Designs 2 and 3 start with a shorter module (10 items), followed by a 15-items module at stage 2, before ending with a longer module (20 items) at stage 3. For designs 3 and 6, module lengths are in reverse order: stage 1 has the largest module (20 items) and stage 3 holds the shorter modules (10 items each). In each of these designs, only 90 items (out of the 100 available items in the 2PL item bank) are considered.

Table 8.1 Module sizes by stage for the six possible MST designs using the 2PL item bank

Design	Stage 1	Stage 2	Stage 3	Total
1	15	15	15	45
2	10	15	20	45
3	20	15	10	45
4	15	15	15	45
5	10	15	20	45
6	20	15	10	45

Table 8.2 Means and variances of item discrimination parameters in each module and for each MST design

	Stage 1	Stage 2		Stage 3		
#	Module A	Module B	Module C	Module D	Module E	Module F
1	0.867 (0.138)	1.221 (0.091)	0.904 (0.094)	1.428 (0.106)	1.016 (0.075)	0.839 (0.049)
2	0.584 (0.040)	1.221 (0.091)	0.904 (0.094)	1.357 (0.099)	1.010 (0.064)	0.789 (0.096)
3	1.001 (0.070)	1.221 (0.091)	0.904 (0.094)	1.526 (0.108)	1.160 (0.072)	0.830 (0.042)
4	0.880 (0.090)	1.250 (0.120)	0.793 (0.046)	1.412 (0.055)	1.182 (0.051)	0.752 (0.085)
5	0.643 (0.030)	1.250 (0.120)	0.793 (0.046)	1.352 (0.066)	1.101 (0.057)	0.734 (0.106)
6	1.055 (0.100)	1.250 (0.120)	0.793 (0.046)	1.415 (0.048)	1.178 (0.029)	0.737 (0.082)

Module names are chosen in agreement with Fig. 7.2

Table 8.3 Means and variances of item difficulty parameters in each module and for each MST design

	Stage 1	Stage 2		Stage 3		
#	Module A	Module B	Module C	Module D	Module E	Module F
1	−0.254 (1.704)	−1.030 (0.351)	0.132 (0.607)	−1.532 (0.173)	−0.497 (0.404)	0.689 (0.400)
2	−0.421 (0.855)	−1.030 (0.351)	0.132 (0.607)	−1.481 (0.176)	−0.575 (0.219)	0.870 (0.965)
3	−0.473 (0.713)	−1.030 (0.351)	0.132 (0.607)	−1.542 (0.220)	−0.493 (0.485)	0.698 (0.504)
4	−0.602 (0.210)	−1.123 (0.152)	0.241 (0.145)	−1.658 (0.101)	−0.611 (0.109)	1.201 (1.027)
5	−0.586 (0.300)	−1.123 (0.152)	0.241 (0.145)	−1.583 (0.120)	−0.563 (0.121)	1.031 (0.956)
6	−0.541 (0.230)	−1.123 (0.152)	0.241 (0.145)	−1.783 (0.090)	−0.641 (0.099)	1.402 (0.775)

Module names are chosen in agreement with Fig. 7.2

In addition to differences in module lengths per stage, the various designs also exhibit differences in item characteristics per module. Tables 8.2 and 8.3 display mean and variance values of item discrimination and difficulty parameters within a module and by test design.

For each MST design, an appropriate module matrix must be created, defining which items belong to which module. The following code describes the creation of the `module1` matrix for MST design 1. Items belonging to a module are recorded (by means of their identification numbers in the item bank) in six vectors, one for each module, labeled `A1` for module A from design 1, `B1` for module B in design 1, and so on. The `module1` matrix is then created.

```
R> A1 <- c(7, 13, 14, 15, 22, 28, 34, 38, 51, 59,
+          70, 84, 85, 87, 89)
R> B1 <- c(4, 12, 19, 27, 29, 39, 43, 60, 69, 71,
+          72, 73, 75, 90, 97)
R> C1 <- c(32, 35, 36, 45, 52, 57, 61, 63, 66,
+          78, 81,82, 88, 92, 100)
R> D1 <- c(1, 2, 3, 6, 18, 21, 41, 48, 49, 50,
+          54, 56, 58, 65, 77)
R> E1 <- c(5, 16, 24, 25, 26, 37, 40, 44, 53, 68,
+          74, 80, 86, 91, 93)
R> F1 <- c(9, 11, 31, 46, 47, 55, 64, 67, 76, 79,
+          83, 94, 95, 98, 99)
R> module1 <- matrix(0, 100, 6)
R> module1[A1, 1] <- module1[B1, 2]
+        <- module1[C1, 3] <- module1[D1, 4]
+        <- module1[E1, 5] <- module1[F1, 6] <- 1
```

The first 6 rows of the module1 matrix are displayed below for direct inspection: items 1–3 belong to module D (the fourth column), item 4 belongs to module B (the second column), and so on.

```
R> head(module1)
     [,1] [,2] [,3] [,4] [,5] [,6]
[1,]    0    0    0    1    0    0
[2,]    0    0    0    1    0    0
[3,]    0    0    0    1    0    0
[4,]    0    1    0    0    0    0
[5,]    0    0    0    0    1    0
[6,]    0    0    0    1    0    0
```

Other modules (labeled module2 for MST design 2 etc.) can also be created according to the particular distribution of items into modules. The full code can be obtained upon request.

Eventually, whatever the design is, the MST structure remained the same (i.e., a 1-2-3 structure as depicted in Fig. 7.2), so a single transition matrix is necessary. The code below allows for creating such a matrix, later referred to as trans.

```
R> trans <- matrix(0, 6, 6)
R> trans[1, 2:3] <- trans[2, 4:5]
+    <- trans[3, 5:6] <- 1
```

The corresponding transition matrix looks as follows:

```
R> trans
        [,1]  [,2]  [,3]  [,4]  [,5]  [,6]
[1,]     0     1     1     0     0     0
[2,]     0     0     0     1     1     0
[3,]     0     0     0     0     1     1
[4,]     0     0     0     0     0     0
[5,]     0     0     0     0     0     0
[6,]     0     0     0     0     0     0
```

With the above settings for design 1, MST simulations using the package **mstR** can be started.

8.2   Example 1: MST Using `randomMST()`

The first example illustrates the basic performance of the `randomMST()` function. The first MST design (characterized by `module1` design matrix) is selected and a random MST pattern is generated for a test taker with true ability level of zero. The stage 1 module is selected as being the most informative one for an ability level of zero. The next module is selected as the most informative one for the current ability estimate by the ML method. Final ability estimation is performed by ML also. Note that although the starting list has to be supplied for compatibility, in this example only module A can be selected as it is the only stage 1 module available in the bank for design 1. But in the scenario of several modules available at stage 1, the most informative starting module would be selected.

The following code creates the three lists of arguments and run the main function with the required specifications. Recall that all arguments of `randomMST()` are available as shown in Table 7.1.

```
R> start <- list(theta = 0)
```

(continued)

```
R> test <- list(method = "ML", moduleSelect = "MFI")
R> final <- list(method = "ML")
R> mst.ex1 <- randomMST(trueTheta = 0,
+              itemBank = it.2PL, modules = module1,
+              transMatrix = trans, genSeed = 1,
+              start = start, test = test,
+              final = final)
```

The full output is displayed below for illustration. Only selected output will be shown for the next examples.

```
R> mst.ex1
Random generation of a MST response pattern
  with random seed equal to 1

  Item bank calibrated under Two-Parameter Logistic
    model

True ability level: 0

MST structure:
  Number of stages: 3
  Structure (number of modules per stage): 1-2-3

Starting parameters:
  Number of available modules at first stage: 1
  Selection of the first stage module: by
  maximizing module information for starting
  ability Starting ability level: 0

Multistage test parameters:
 Next module selection: maximum Fisher information
 Provisional ability estimator: Maximum likelihood
                                estimator
 Provisional range of ability values: [-4,4]
 Ability estimation adjustment for constant
   pattern: none

Multistage test details:
```

(continued)

```
Stage 1:
  Module administered: 1
  Number of items in module 1: 15 items
  Items and responses:

Nr      1  2  3  4  5  6  7  8  9 10 11 12 13 14 15
Item    7 13 14 15 22 28 34 38 51 59 70 84 85 87 89
Resp.   1  0  1  0  1  0  1  1  1  1  1  0  0  1  1

  Provisional ability estimate (SE) after stage 1:
    0.204 (0.629)

Stage 2:
  Module administered: 2
  Number of items in module 2: 15 items
  Items and responses:

Nr      1  2  3  4  5  6  7  8  9 10 11 12 13 14 15
Item    4 12 19 27 29 39 43 60 69 71 72 73 75 90 97
Resp.   1  1  1  0  1  0  0  1  1  1  1  1  1  0  1

  Provisional ability estimate (SE) after stage 2:
    -0.062 (0.397)

Stage 3:
  Module administered: 5
  Number of items in module 5: 15 items
  Items and responses:

Nr      1  2  3  4  5  6  7  8  9 10 11 12 13 14 15
Item    5 16 24 25 26 37 40 44 53 68 74 80 86 91 93
Resp.   1  1  1  1  1  0  0  1  1  1  1  1  1  0  1

  Provisional ability estimate (SE) after stage 3:
    0.152 (0.334)

Final results:
  Total length of multistage test: 45 items
  Final ability estimator: Maximum likelihood
                           estimator
  Final range of ability values: [-4,4]
```

(continued)

```
Final ability estimate (SE): 0.152 (0.334)
95% confidence interval: [-0.502,0.806]

Output was not captured!
```

The output is rather similar to that of the randomCAT() function as displayed in Chap. 5. The first part of the output recapitulates the design of the MST and the various options chosen to start the MST, followed by the next stage module selection and the final results. The details of each stage are displayed in subsequent tables with relevant results: the selected module and the number of items per module, the generated response pattern, and the provisional ability estimate and related standard error. Eventually, final test length, ability estimate, standard error, and confidence interval (based on a normal distribution assumption) are returned.

In this simulated example, responses from module A from stage 1 yield a provisional ability estimate of 0.204. For this value, the most informative module from stage 2 is module B, as shown in the output. Note that this can be inspected with the following code:

```
R> sum(Ii(0.204, it.2PL[module1[, 2] == 1, ])$Ii)
[1] 2.946685
R> sum(Ii(0.204, it.2PL[module1[, 3] == 1, ])$Ii)
[1] 2.791439
```

When all items from module B are administered, the 30 provisional responses from both stages yield a provisional estimate of −0.062. With this value, the next module is chosen to be module E, which can also be confirmed using code similar to the one displayed above. After all 45 items are administered, the final ability estimate equals 0.152 with standard error 0.334. The 95% final confidence interval covers the true ability level (set as zero), indicating that true ability level is accurately covered.

The resulting MST plot and generated path can be displayed using the following code. By default, the path is displayed in red color, both for the modules and the connecting paths, but for ease of presentation in this book it is displayed in grey color. The graphical output is shown in Fig. 8.1; further graphical options are available, as shown in Table 7.4.

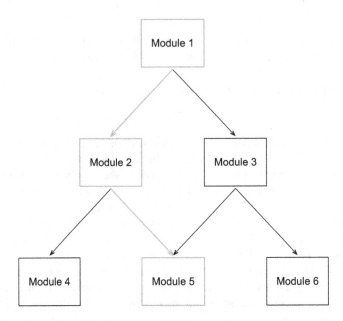

Fig. 8.1 Graphical output of the first simulated MST example

```
R> plot(mst.ex1, border.col = "grey",
+           arrow.col = "grey")
```

8.3   Example 2: MST with Cut-Scores

One advantage of MST (with respect to CAT) is that thresholds can be pre-specified across modules within the stage. In this section we illustrate how to specify such thresholds for the next module selection, first by using ability estimation, then by using test score. To simplify the example, the same general MST design and options of Sect. 8.2 will be considered whenever possible.

8.3.1   Thresholds for Ability Estimation

The selection of the modules is based on the following ability thresholds:

- at stage 2, module B is chosen if the provisional ability estimate is smaller than or equal to zero, and module C otherwise;

- at stage 3, module D is chosen if the provisional ability estimate is smaller than or equal to -0.3, module F is chosen if provisional ability estimate is larger than 0.3, and module E otherwise.

As described in Sect. 7.3, these thresholds can be set in **mstR** by designing the following cut matrix:

```
R> cut <- matrix(NA, 6, 2)
R> cut[2,] <- c(-Inf, 0)
R> cut[3,] <- c(0, Inf)
R> cut[4,] <- c(-Inf,-0.3)
R> cut[5,] <- c(-0.3, 0.3)
R> cut[6,] <- c(0.3, Inf)
```

To incorporate this threshold matrix for selecting the next module, the following list is created:

```
R> test2 <- list(method = "ML", cutoff = cut)
```

Note that defining the ML estimator is mandatory since the default estimator is BM. However no moduleSelect value must be provided since module selection will be based on the thresholds defined above.

The random MST generation is now run with the following code (note that true ability level was increased to 1 in this example):

```
R> mst.ex2 <- randomMST(trueTheta = 1,
+              itemBank = it.2PL, modules = module1,
+              transMatrix = trans, genSeed = 1,
+              start = start, test = test2,
+              final = final)
```

Selected parts of the output are displayed hereafter.

```
Random generation of a MST response pattern
  with random seed equal to 1

 (...)
```

(continued)

```
     Multistage test parameters:
       Next module selection: by pre-specified cut
       scores  (random selection among allowed modules)

  (...)

  Multistage test details:

     Stage 1:
       Module administered: 1

  (...)

       Provisional ability estimate (SE) after stage 1:
         0.64 (0.704)

     Stage 2:
       Module administered: 3

  (...)

       Provisional ability estimate (SE) after stage 2:
         1.034 (0.524)

     Stage 3:
       Module administered: 6

  (...)

       Provisional ability estimate (SE) after stage 3:
         0.953 (0.398)

     Final results:

  (...)

       Final ability estimate (SE): 0.953 (0.398)
       95% confidence interval: [0.174,1.733]
```

After stage 1, the provisional ability estimate is 0.64 and is larger than the stage
2 threshold (set as 0.3), so module C is selected. After stage 2 administration, the
provisional ability estimate increases to 1.034, leading to selection of module F at

stage 3 (since the largest threshold is 0.3 at stage 3). The final ML ability estimate
is 0.953, very close to the true level of 1.

8.3.2 Score-Based Thresholds

Instead of IRT ability estimates, score-based rules are now used to select the next
module and to evaluate ability based on test scores. Since in this design each
stage provides 15 additional items, the following thresholds (chosen arbitrarily) are
considered:

- at stage 2, module B is chosen if the provisional test score is smaller than or equal
  to 7, and module C otherwise;
- at stage 3, module D is chosen if the provisional ability estimate is smaller than or
  equal to 9, module E if provisional test score lies between 10 and 20, and module
  F is chosen if the provisional test score is larger to 20.

To avoid possible confusion with integer thresholds, it is preferable (but not
mandatory) to make use of real values in the cut-off matrix. For instance, threshold
7.5 at stage 2 defines the set of scores (less than or equal to 7, and larger than 7) for
each module. Thus, a similar cut-off matrix cut2 is defined, now with appropriate
values:

```
R> cut2 <- matrix(NA, 6, 2)
R> cut2[2,] <- c(-Inf, 7.5)
R> cut2[3,] <- c(7.5, Inf)
R> cut2[4,] <- c(-Inf,9.5)
R> cut2[5,] <- c(9.5, 20.5)
R> cut2[6,] <- c(20.5, Inf)
```

This threshold matrix is used as input for the next module selection and the R
code for the two lists looks like this:

```
R> test3 <- list(method = "score", cutoff = cut2)
R> final3 <- list(method = "score")
```

Setting "score" as a value for the method argument yields test scores instead
of IRT ability estimates.

The random MST generation is now run with the following code, similarly to the
previous example:

```
R> mst.ex3 <- randomMST(trueTheta = 1,
+              itemBank = it.2PL, modules = module1,
+              transMatrix = trans, genSeed = 1,
+              start = start, test = test3,
+              final = final)
```

Selected output is displayed hereafter.

```
> ess3
Random generation of a MST response pattern
  with random seed equal to 1

(...)

  Multistage test parameters:
    Next module selection: by pre-specified cut
    scores (random selection among allowed modules)
    Provisional ability estimator: Test score
        (sum-score) computation
    Ability estimation adjustment for constant
        pattern: none

  Multistage test details:

    Stage 1:
      Module administered: 1

(...)

      Provisional ability estimate (SE) after stage 1:
        11 (NA)

    Stage 2:
      Module administered: 3

(...)

      Provisional ability estimate (SE) after stage 2:
        22 (NA)
```

(continued)

```
    Stage 3:
      Module administered: 6

(...)

      Provisional ability estimate (SE) after stage 3:
        30 (NA)
    Final results:
      Total length of multistage test: 45 items
      Final ability estimator: Total test score
      Final ability estimate (SE): 30 (NA)
```

All results are logical with respect to the fixed cutoff scores set in the cut2 matrix. Intermediate and final results only show the test scores. Note that the SE is obviously not computed for test scores.

8.4   Example 3: Comparing MST Designs

More elaborated simulations are now displayed. In this section, various MST designs (depicted by the six possible designs described in Table 8.1) are compared. The design of the study is displayed first, followed by the appropriate R code. The output is then further analyzed.

8.4.1   Designs

The aim of this example is to investigate the potential effect of various MST configurations on the final estimation of abilities using IRT models. The six aforementioned designs will be compared. Since they all share the same underlying structure (that is, a 1-2-3 design), only the impact of module length across stages, as well as the module characteristics themselves, will be considered as potential sources of variation in the final IRT ability estimators. All options for module selection and ability estimation will be kept constant in all designs.

More precisely, the following study is considered. One set of true ability levels is created, by replicating 1000 times each value from −2 to 2 by steps of one half (that is, 9000 true levels in total). For each ability level, one MST pattern was generated for each of the six MST designs, using the same item bank (the 2PL one). The following options were considered for the 54,000 generated patterns:

1. The first module was chosen as the most informative one with starting ability level zero.
2. The next module is chosen by maximizing its Fisher information for the current ability level.
3. Throughout the MST, ability is estimated using the BM method with a standard normal prior distribution.
4. Final ML estimation is returned at the end of the MST.

Each final ability estimate is recorded, and both ASB and RMSE values are computed by true ability level and by MST design for further comparisons.

8.4.2 Implementation

The code displayed below describes how the example design can be implemented to extract all required ability estimates. First, the sequence of true ability levels TH is created by replicating the regular sequence s of ability levels. Second, the result matrix RES with as many rows as the number of true ability levels (i.e., 9000) and six columns (one for each MST design) is built. Third, the three lists of options are defined.

```
s <- seq(-2, 2, 0.5)
TH <- rep(s, each = 1000)
RES <- matrix(NA, length(TH), 6)
start <- list(theta = 0)
test <- list(method = "BM", moduleSelect = "MFI")
final <- list(method = "ML")
```

Now, the following loops across all MST designs and all true ability levels are built. At each iteration, an MST pattern is drawn with the randomMST() and the final ability estimate is stored at its appropriate location in the RES matrix.

```
R> for (j in 1:6) {
+   mod <- get(paste("module", j, sep = ""))
+   for (i in 1:length(TH)){
+   prov <- randomMST(trueTheta = TH[i],
+   itemBank = it.2PL, modules = mod,
+   transMatrix = trans, genSeed = i, start = start,
+   test = test, final = final)
```

(continued)

```
+  RES[i, j] <- prov$thFinal
+  }}
```

In sum, the RES matrix contains the final MST ability estimates for each generated true ability level (per row) and each MST design (per column). This matrix, together with the true generated abilities, is available in a csv file upon request.

8.4.3 Results

Making use of the vector TH of true ability levels, one can compute ASB and RMSE values per design and true level, as follows. First, the output matrices are created, one for ASB values (res.ASB) and one for RMSE values (res.RMSE). The number of rows equals the number of unique true ability levels. Second, two functions called ASB() and RMSE() are created to compute the ASB and RMSE values over a set of estimated abilities and given true ability levels.

```
R> res.ASB <- res.RMSE <- matrix(NA, length(s), 6)
R> ASB <- function(t, th) mean(t-th)
R> RMSE <- function(t, th) sqrt(mean((t-th)^2))
```

Then, these functions are applied repeatedly across all designs and true ability levels to fill in the output tables with summary statistics:

```
R> for (i in 1:length(s)){
+  for (j in 1:6){
+  ind <- which(TH == s[i])
+  res.ASB[i, j] <- ASB(RES[ind, j], s[i])
+  res.RMSE[i, j] <- RMSE(RES[ind, j], s[i])
+  }}
```

Summary statistics can now be displayed graphically, as in Fig. 8.2. The left panel displays the ASB values per true ability level and MST design, while the right panel displays the corresponding RMSE values.

The primary conclusions from Fig. 8.2 is that none of the MST design clearly under-performs others nor outperforms others. Some variation among the designs is

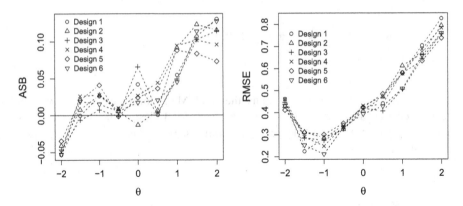

Fig. 8.2 ASB (left) and RMSE (right) values for each MST design and true ability level generated in Example 3

clearly present, both for bias and RMSE values, but all six designs have a similar pattern: slight underestimation for low ability levels, slight overestimation for large ability levels, and increased general variability at the extremes of the ability scale. The smallest RMSEs are observed for ability levels around -1, which is similar to what is observed under CAT scenarios (e.g., Sect. 5.1.1).

In summary, with respect to final ability estimation, this small simulation example (with fixed MST parameters and various designs) tends to highlight that neither the sizes of the modules at each stage, nor the composition of these modules (as they were tested here) significantly affect the ability estimation process. Many other aspects could be investigated, for instance the variability in routing results, the frequency of chosen modules, as well as the stability of these results with respect to other MST parameters. Those aspects can easily be simulated and analyzed with a similar approach using **mstR** functions as described in this section.

8.5 Example 4: MST Versus CAT

To end up this illustrative chapter, we provide a simulated comparison of MST administration versus CAT administration of the items from the 2PL bank. The following example compares final ability estimates under these two adaptive designs, with options chosen as similar as possible. Post-hoc simulations are considered to make use of exactly the same item responses in both designs.

8.5.1   Design and Code

As in the previous example, 9000 true ability levels are considered, namely 1000 repetitions of each value in the sequence from -2 to 2 by steps of 0.5. For each true ability level (generated and stored in the TH vector), a response pattern is drawn on the whole 2PL item bank, so that post-hoc simulations in CAT and MST will be run, based on the same response patterns. The matrix DATA holds these response patterns, generated using the genPattern() function. Moreover, the output values to be stored are the final ability estimates and related standard errors, in both the CAT and MST designs. The output matrix RES is therefore created (first as an empty matrix) with 9000 rows and four columns, to contain the final CAT estimate, the final CAT associated SE, the final MST estimate and the final MST associated SE, respectively.

```
R> s <- seq(-2, 2, 0.5)
R> TH <- rep(s, each = 1000)
R> DATA <- genPattern(TH, it.2PL, seed = 1)
R> RES <- matrix(NA, length(TH), 4)
```

Now, CAT and MST options are defined. First, the test starts by selecting the most informative item (in CAT) or module (in MST) for a starting ability level of zero. In CAT simulations, the "randomesque" selection among the top five items is also required. Second, provisional ability estimation is performed by the BM approach with the standard normal prior distribution. Next item (in CAT) and module (in MST) selection is performed by maximum Fisher (item or module) information. In CAT scenarios, randomesque selection among the five most informative items is performed. Third, the CAT stopping rule coincides with the MST by-default rule, that is, the CAT stops once 45 items are administered (so that all CAT and MST patterns will have exactly the same length). Eventually, final ability estimation is performed by BM with standard normal prior distributions in both designs.

These options are implemented with the following code. Note that the final list is common to both designs, while starting and testing lists must be split into two versions given that randomesque selection in CAT must be specified but not in MST. Moreover, the stop list is useful only in CAT.

```
R> start.mst <- list(theta = 0)
R> start.cat <- list(theta = 0, randomesque = 5)
R> test.cat <- list(method = "BM",
+                   itemSelect = "MFI",
```

(continued)

```
+                              randomesque = 5)
R> test.mst <- list(method = "BM",
+                              moduleSelect = "MFI")
R> stop <- list(rule = "length", thr = 45)
R> final <- list(method = "BM")
```

Eventually, two response patterns are drawn for each test taker, one under the CAT design and one under the MST design, using the post-hoc simulation scenario. In the following loop, the CAT output is first generated with the `randomCAT()` function (and stored in output list `pr`), and the final ability estimate (`pr$thFinal`) and related standard error (`pr$seFinal`) are extracted and saved in the `RES` table. Similar actions are taken for the MST design but with the `randomMST()` function instead.

```
for (i in 1:length(TH)){
pr <- randomCAT(trueTheta = TH[i],
itemBank = it.2PL, genSeed = i, start = start.cat,
 test = test.cat, stop = stop, final = final)
RES[i,1] <- pr$thFinal
RES[i,2] <- pr$seFinal
pr <- randomMST(trueTheta = TH[i],
itemBank = it.2PL, genSeed = i, modules = module1,
transMatrix = trans, start = start.mst,
test = test.mst, final = final)
RES[i,3] <- pr$thFinal
RES[i,4] <- pr$seFinal
}
```

8.5.2 Results

From the `RES` output table, three summary statistics are computed at each true ability level and for each design: the average signed bias, the average final SE of ability, and the RMSE values. They are displayed in Fig. 8.3, respectively, in upper right, lower left, and lower right panels. The upper left panel displays the scatter plot of CAT versus MST ability estimates for illustration purposes. Note that the sample correlation between the two ability estimates is quite large (0.984), indicating rather coherent final estimates with the two designs.

Before discussing these results further, it is noticeable that the bias curves in the upper right panel of Fig. 8.3 have the opposite trend of those displayed in left panel

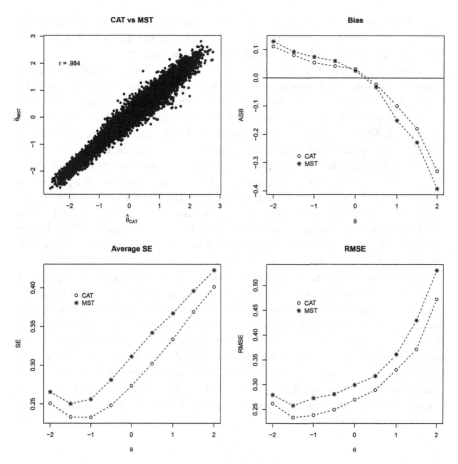

Fig. 8.3 Scatter plot of CAT versus MST ability estimates (upper left), ASB (upper right), average SE (lower left) and RMSE (lower right) values for CAT and MST designs at each true ability level generated in Example 4

of Fig. 8.2 in the previous section. However, the same item bank and MST design 1 were chosen, as well as most options for adaptive designs. The main difference is actually in the final ability estimator, chosen as the ML estimator in Sect. 8.4 and as the BM estimator in this section. It was already established (Lord, 1983, 1986) that with the ML estimator, very small ability levels are underestimated and very large ability levels are overestimated. With Bayesian estimators (such as BM) however, these trends are reversed due to the shrinkage of ability estimates towards the center of the prior distribution. In other words, these opposite trends can be explained by the choice of the final ability estimator.

The output from the panels of Fig. 8.3 are insightful in three main directions. First, globally both CAT and MST designs return ability estimates that behave similarly: they are more biased at the extremes of the scales (positively biased at lower levels and negatively biased at higher levels) and more precise for ability levels around -1 (as was pointed out in Chap. 5). The two adaptive designs yield similar trends, which is the first remarkable fact to highlight.

Second, and as could be anticipated, the CAT design is more accurate in ability estimation than the MST approach. The bias is always smaller (in absolute value) for CAT estimates than for MST estimates, while both the average SE and the RMSE are consistently smaller for CAT simulations across the whole ability range. Since the aim of CAT is to optimize the administration of each item, it is expected that such item-by-item optimization yields final results that are more precise and less biased than module-by-module administration, especially at the extremes of the ability scale. In other words, despite their intrinsic characteristics and the same philosophy in selecting the modules, some items within a module are not optimal with respect to the true ability level to be estimated, as they are in a CAT scenario. Of course MST has some practical advantages over CAT (some were listed in Chap. 6), but for the single purpose of point ability estimation, CAT remains the most accurate method (with all other things being equal).

Third, even though the CAT design seems more accurate in terms of precision in ability estimation, the gap between CAT and MST designs is not that large. For instance, the maximum difference in bias is about 0.06 and is observed at the largest true ability levels. Similarly, the largest difference in RMSE values is smaller than 0.06. In other words, the optimality of individual item administration (in a CAT scenario) with respect to the block-item (i.e., module) administration (in a MST design) has a clear and consistent impact on estimation bias and variability, but remains limited for tests of 45 items. This indicates indirectly that the MST design is rather adequate for accurately estimating ability levels with a 1-2-3 design and for the chosen built-in modules.

References

Ackerman, T. A. (1989). Unidimensional IRT calibration of compensatory and non-compensatory multidimensional items. *Applied Psychological Measurement, 13*, 113–127. https://doi.org/10.1177/014662168901300201

Adams, R. J., Wilson, M., & Wang, W. (1997). The multidimensional random coefficients multinomial logit model. *Applied Psychological Measurement, 21*, 1–24. https://doi.org/10.1177/0146621697211001

Andersen, E. B. (1970). Asymptotic properties of conditional maximum likelihood equations. *Journal of the Royal Statistical Society, Series B, 32*, 283–301.

Andrich, D. (1978). A rating formulation for ordered response categories. *Psychometrika, 43*, 561–573. https://doi.org/10.1007/BF02293814

Angoff, W. H., & Huddleston, E. M. (1958). *The multi-level experiment: A study of a two-level test system for the college board scholastic aptitude test* (ETS Statistical Report No. SR-58-21). Princeton, NJ: Educational Testing Service.

Baker, F. B., & Kim, S.-H. (2004). *Item response theory: Parameter estimation techniques* (2nd ed.). New York: Marcel Dekker.

Barrada, J. R., Abad, F. J., & Veldkamp, B. P. (2009). Comparison of methods for controlling maximum exposure rates in computerized adaptive testing. *Psicothema, 21*, 313–320.

Barrada, J. R., Mazuela, P., & Olea, J. (2006). Maximum information stratification method for controlling item exposure in computerized adaptive testing. *Psicothema, 18*, 156–159.

Barrada, J. R., Olea, J., Ponsoda, V., & Abad, F. J. (2008). Incorporating randomness to the fisher information for improving item exposure control in cats. *British Journal of Mathematical and Statistical Psychology, 61*, 493–513. https://doi.org/10.1348/000711007X230937

Barrada, J. R., Olea, J., Ponsoda, V., & Abad, F. J. (2009). Item selection rules in computerized adaptive testing: Accuracy and security. *Methodology, 5*, 7–17. https://doi.org/10.1027/1614-2241.5.1.7

Barrada, J. R., Olea, J., Ponsoda, V., & Abad, F. J. (2010). A method for the comparison of item selection rules in computerized adaptive testing. *Applied Psychological Measurement, 34*, 438–452. https://doi.org/10.1177/0146621610370152

Barton, M. A., & Lord, F. M. (1981). *An upper asymptote for the three-parameter logistic item-response model* (Research Bulletin No. 81-20). Princeton, NJ: Educational Testing Service.

Bejar, I. I., Lawless, R. R., Morley, M. E., Wagner, M. E., Bennett, R. E., & Revuelta, J. (2003). A feasibility study of on-the-fly item generation in adaptive testing. *Journal of Technology, Learning, and Assessment, 2*(3), 3–29.

© Springer International Publishing AG 2017

D. Magis et al., *Computerized Adaptive and Multistage Testing with R*, Use R!,

https://doi.org/10.1007/978-3-319-69218-0

Belov, D. I., & Armstrong, R. D. (2009). Direct and inverse problems of item pool design for computerized adaptive testing. *Educational and Psychological Measurement, 69*, 533–547. https://doi.org/10.1177/0013164409332224

Betz, N. E., & Weiss, D. J. (1974). *Simulation studies of two-stage ability testing* (Research Report 74-4). Minneapolis: University of Minnesota.

Birnbaum, A. (1968). Some latent trait models and their use in inferring an examinee's ability. In F. M. Lord & M. R. Novick (Eds.), *Statistical theories of mental test scores*. Reading, MA: Addison-Wesley.

Birnbaum, A. (1969). Statistical theory for logistic mental test models with a prior distribution of ability. *Journal of Mathematical Psychology, 6*, 258–276. https://doi.org/10.1016/0022-2496(69)90005-4

Bock, R. D. (1972). Estimating item parameters and latent ability when responses are scored in two or more nominal categories. *Psychometrika, 37*, 29–51. https://doi.org/10.1007/BF02291411

Bock, R. D., & Aitkin, M. (1981). Marginal maximum likelihood estimation of item parameters: Application of an EM algorithm. *Psychometrika, 46*, 443–459. https://doi.org/10.1007/BF02293801

Bock, R. D., & Lieberman, M. (1970). Fitting a response model for n dichotomously scored items. *Psychometrika, 35*, 179–197. https://doi.org/10.1007/BF02291262

Bock, R. D., & Mislevy, R. J. (1982). Adaptive EAP estimation of ability in a microcomputer environment. *Applied Psychological Measurement, 6*, 431–444. https://doi.org/10.1177/014662168200600405

Bradlow, E. T. (1996). Negative information and the three-parameter logistic model. *Journal of Educational and Behavioral Statistics, 21*, 179–185. https://doi.org/10.2307/1165216

Braeken, J., Tuerlinckx, F., & De Boeck, P. (2007). Copulas for residual dependencies. *Psychometrika, 72*, 393–411. https://doi.org/10.1007/s11336-007-9005-4

Breiman, L., Friedman, L., Stone, C. J., & Olshen, R. A. (1984). *Classification and regression trees*. New York: CRC Press.

Chalmers, R. P. (2016). Generating adaptive and non-adaptive test interfaces for multidimensional item response theory applications. *Journal of Statistical Software, 71*(5), 1–39. https://doi.org/10.18637/jss.v071.i05

Chang, H.-H., & Ying, Z. (1996). A global information approach to computerized adaptive testing. *Applied Psychological Measurement, 20*, 213–229. https://doi.org/10.1177/014662169602000303

Chang, S., & Ansley, T. (2003). A comparative study of item exposure control methods in computerized adaptive testing. *Journal of Educational Measurement, 40*, 71–103. https://doi.org/10.1111/j.1745-3984.2003.tb01097.x

Chen, W.-H., & Thissen, D. (1997). Local dependence indexes for item pairs using item response theory. *Journal of Educational and Behavioral Statistics, 22*, 265–289. https://doi.org/10.3102/10769986022003265

Cheng, Y. (2009). When cognitive diagnosis meets computerized adaptive testing: CD-CAT. *Psychometrika, 74*, 619–632. https://doi.org/10.1007/s11336-009-9123-2

Cheng, Y., & Chang, H.-H. (2009). The maximum priority index method for severely constrained item selection in computerized adaptive testing. *British Journal of Mathematical and Statistical Psychology, 62*, 369–383. https://doi.org/10.1348/000711008X304376

Cheng, Y., Chang, H.-H., Douglas, J., & Guo, F. (2009). Constraint-weighted a-stratification for computerized adaptive testing with nonstatistical constraints: Balancing measurement efficiency and exposure control. *Educational and Psychological Measurement, 69*, 35–49. https://doi.org/10.1177/0013164408322030

Choi, S. W. (2009). FIRESTAR: Computerized adaptive testing simulation program for polytomous item response theory models. *Applied Psychological Measurement, 33*, 644–645. https://doi.org/10.1177/0146621608329892

Choi, S. W., & Swartz, R. J. (2009). Comparison of CAT item selection criteria for polytomous items. *Applied Psychological Measurement, 32*, 419–440. https://doi.org/10.1177/0013164408322030

Cronbach, L. J., & Gleser, G. C. (1965). *Psychological tests and personnel decisions*. Urbana, IL: University of Illinois Press.

Davey, T., & Parshall, C. G. (1999). *New algorithms for item selection and exposure control with computerized adaptive testing*. Paper presented at the annual meeting of the American Educational Research Association.

De Boeck, P., & Wilson, M. (2004). *Explanatory item response models: A generalized linear and nonlinear approach*. New York: Springer.

DeMars, C. (2010). *Item response theory*. Oxford: Oxford University Press.

Dodd, B. G., De Ayala, R. J., & Koch, W. R. (1995). Computerized adaptive testing with polytomous items. *Applied Psychological Measurement, 19*, 5–22. https://doi.org/10.1177/014662169501900103

Drasgow, F., Levine, M. V., & Williams, E. A. (1985). Appropriateness measurement with poly-chotomous item response models and standardized indices. *British Journal of Mathematical and Statistical Psychology, 38*, 67–86. https://doi.org/10.1111/j.2044-8317.1985.tb00817.x

Eggen, T. J. H. M. (2010). Three-category adaptive classification testing. In W. V. der Linden & C. A. W. Glas (Eds.), *Elements of adaptive testing* (pp. 373–387). New York: Springer.

Eggen, T. J. H. M., & Verhelst, N. D. (2011). Item calibration in incomplete testing designs. *Psicologica, 32*, 107–132.

Embretson, S. E., & Reise, S. P. (2000). *Item response theory for psychologists*. Mahwah, NJ: Lawrence Erlbaum Associates.

Finch, H., & Habing, B. (2007). Performance of DIMTEST- and NOHARM-based statistics for testing unidimensionality. *Applied Psychological Measurement, 31*, 292–307. https://doi.org/10.1177/0146621606294490

Fischer, G. H. (1981). On the existence and uniqueness of maximum-likelihood estimates in the Rasch model. *Psychometrika, 46*, 59–77. https://doi.org/10.1007/BF02293919

Fraser, C., & McDonald, R. P. (2003). *Noharm 3.0 [Computer software manual]*. http://people.niagaracollege.ca/cfraser/download/

Geerlings, H., Glas, C., & van der Linden, W. J. (2011). Modeling rule-based item generation. *Psychometrika, 76*, 337–359. https://doi.org/10.1007/s11336-011-9204-x

Gessaroli, M. E., & De Champlain, A. F. (1996). Using a approximate chi-square statistic to test the number of dimensions underlying the responses to a set of items. *Journal of Educational Measurement, 33*, 157–179. https://doi.org/10.1111/j.1745-3984.1996.tb00487.x

Glas, C. A. W., & van der Linden, W. J. (2003). Computerized adaptive testing with item cloning. *Applied Psychological Measurement, 27*, 247–261. https://doi.org/10.1177/0146621603027004001

Glas, C. A. W., & Vos, H. J. (2010). Adaptive mastery testing using a multidimensional IRT model. In W. V. der Linden & C. A. W. Glas (Eds.), *Elements of adaptive testing* (pp. 409–431). New York: Springer.

Green, B. F. J. (1950). *A general solution for the latent class model of latent structure analysis* (ETS Research Bulletin Series No. RB-50-38). Princeton, NJ: Educational Testing Service.

Haberman, S. J., & von Davier, A. A. (2014). Considerations on parameter estimation, scoring, and linking in multistage testing. In D. Yan, A. A. von Davier, & C. Lewis (Eds.), *Computerized multistage testing: Theory and applications* (pp. 229–248). New York: CRC Press.

Haebara, T. (1980). Equating logistic ability scales by a weighted least squares method. *Japanese psychological Research, 22*, 144–149.

Haley, D. (1952). *Estimation of the dosage mortality relationship when the dose is subject to error* (Technical report No. 15). Palo Alto, CA: Applied Mathematics and Statistics Laboratory, Stanford University.

Hambleton, R. K., & Swaminathan, H. (1985). *Item response theory: Principles and applications*. Boston: Kluwer.

Hambleton, R. K., & Zenisky, A. L. (2013). Reporting test scores in more meaningful ways: A research-based approach to score report design. In K. F. Geisinger, et al. (Eds.), *APA handbook of testing and assessment in psychology, Vol. 3. Testing and assessment in school psychology and education* (pp. 479–494). Washington, DC: American Psychological Association. https://doi.org/10.1037/14049-023

Hattie, J. (1984). An empirical study of various indices for determining unidimensionality. *Multivariate Behavioral Research, 19*, 49–78. https://doi.org/10.1207/s15327906mbr1901_3

Hendrickson, A. (2007). An NCME instructional module on multistage testing. *Educational Measure: Issues and Practice, 26*, 44–52. https://doi.org/10.1111/j.1745-3992.2007.00093.x

Hetter, R. D., & Sympson, J. B. (1997). Item exposure control in CAT-ASVAB. In J. R. McBride (Ed.), *Computerized adaptive testing: From inquiry to operation* (pp. 141–144). Washington, D.C.: American Psychological Association.

Holland, P. W. (1990). On the sampling theory foundations of item response theory models. *Psychometrika, 55*, 577–602. https://doi.org/10.1007/BF02294609

Holland, P. W., & Thayer, D. T. (1988). Differential item performance and the Mantel–Haenszel procedure. In H. Wainer & H. I. Braun (Eds.), *Test validity* (pp. 129–145). Hillsdale, NJ: Erlbaum.

Hsu, C.-L., Wang, W.-C., & Chen, S.-Y. (2013). Variable-length computerized adaptive testing based on cognitive diagnosis models. *Applied Psychological Measurement, 37*, 563–582. https://doi.org/10.1177/0146621613488642

Huitzing, H. A., Veldkamp, B. P., & Verschoor, A. J. (2005). Infeasibility in automated test assembly models: A comparison study of different methods. *Journal of Educational Measurement, 42*, 223–243. https://doi.org/10.1111/j.1745-3984.2005.00012.x

Irvine, S., & Kyllonen, P. (2002). *Item generation for test development*. Mahwah, NJ: Lawrence Erlbaum

Jeffreys, H. (1939). *Theory of probability*. Oxford, UK: Oxford University Press.

Jeffreys, H. (1946). An invariant form for the prior probability in estimation problems. *Proceedings of the Royal Society of London. Series A, Mathematical and Physical Sciences, 186*, 453–461.

Kaplan, M., de la Torre, J., & Barrada, J. R. (2015). New item selection methods for cognitive diagnosis computerized adaptive testing. *Applied Psychological Measurement, 39*, 167–188. https://doi.org/10.1177/0146621614554650

Karabatsos, G. (2003). Comparing the aberrant response detection performance of thirty six person-fit statistics. *Applied Measurement in Education, 16*, 277–298. https://doi.org/10.1207/S15324818AME1604_2

Kelderman, H., & Rijkes, C. P. M. (1994). Loglinear multidimensional IRT models for polytomously scored items. *Psychometrika, 59*, 149–176. https://doi.org/10.1007/BF02295181

Kingsbury, G. G., & Zara, A. R. (1989). Procedures for selecting items for computerized adaptive tests. *Applied Measurement in Education, 2*, 359–375. https://doi.org/10.1207/s15324818ame0204_6

Kingsbury, G. G., & Zara, A. R. (1991). A comparison of procedures for content-sensitive item selection in computerized adaptive tests. *Applied Measurement in Education, 4*, 241–261. https://doi.org/10.1207/s15324818ame0403_4

Klein Entink, R. H., Fox, J.-P., & van der Linden, W. J. (2009). A multivariate multilevel approach to the modeling of accuracy and speed of test takers. *Psychometrika, 74*, 21–48. https://doi.org/10.1007/s11336-008-9075-y

Kosinski, M., Lis, P., Mahalingam, V., Kielczewski, B., Okubo, T., Stillwell, D., et al. (2013). *Concerto - Open-source online R-based adaptive testing platform (Version 4.0) [Computer software manual]*. Cambridge, UK: The Psychometrics Centre.

Leung, C. K., Chang, H.-H., & Hau, K. T. (2003). Computerized adaptive testing: A comparison of three content balancing methods. *The Journal of Technology, Learning and Assessment, 2*, 1–15.

Lewis, C., & Sheehan, K. (1990). Using Bayesian decision theory to design a computerized mastery test. *Applied Psychological Measurement, 14*, 367–386. https://doi.org/10.1177/014662169001400404

Linn, R. L., Rock, D. A., & Cleary, T. A. (1968). *The development and evaluation of several programmed testing methods* (ETS Research Bulletin Series No. i-29). Princeton, NJ: Educational Testing Service.

Lord, F. M. (1951). *A theory of test scores and their relation to the trait measured* (ETS Research Bulletin Series No. RB-51-13). Princeton, NJ: Educational Testing Service.

Lord, F. M. (1971a). The self-scoring flexilevel test. *Journal of Educational Measurement, 8*, 147–151. https://doi.org/10.1111/j.1745-3984.1971.tb00918.x

Lord, F. M. (1971b). A theoretical study of two-stage testing. *Psychometrika, 36*, 227–242. https://doi.org/10.1007/BF02297844

Lord, F. M. (1974). *Practical methods for redesigning a homogeneous test, also for designing a multilevel test* (Research Bulletin No. RB-74-30). Princeton, NJ: Educational Testing Service.

Lord, F. M. (1977). A broad-range tailored test of verbal ability. *Applied Psychological Measurement, 1*, 95–100. https://doi.org/10.1177/014662167700100115

Lord, F. M. (1980). *Applications of item response theory to practical testing problems*. Hillsdale, NJ: Erlbaum.

Lord, F. M. (1983). Unbiased estimators of ability parameters, of their variance, and of their parallel-forms reliability. *Psychometrika, 48*, 233–245. https://doi.org/10.1007/BF02294018

Lord, F. M. (1986). Maximum likelihood and bayesian parameter estimation in item response theory. *Journal of Educational Measurement, 23*, 157–162. https://doi.org/10.1111/j.1745-3984.1986.tb00241.x

Lord, F. M., & Novick, M. R. (1968). *Statistical theories of mental test scores*. Reading, MA: Addison-Wesley.

Luecht, R. M. (1998). Computer-assisted test assembly using optimization heuristics. *Applied Psychological Measurement, 22*, 224–236. https://doi.org/10.1177/01466216980223003

Luecht, R. M. (2014). Systems, design and implementation of large-scale multistage testing. In D. Yan, A. A. von Davier, & C. Lewis (Eds.), *Computerized multistage testing: Theory and applications* (pp. 69–83). New York: CRC Press.

Luecht, R. M., & Nungester, R. J. (1998). Some practical examples of computer-adaptive sequential testing. *Journal of Educational Measurement, 35*, 229–249. https://doi.org/10.1111/j.1745-3984.1998.tb00537.x

Magis, D. (2013). A note on the item information function of the four-parameter logistic model. *Applied Psychological Measurement, 37*, 304–315. https://doi.org/10.1177/0146621613475471

Magis, D. (2014). On the asymptotic standard error of a class of robust estimators of ability in dichotomous item response models. *British Journal of Mathematical and Statistical Psychology, 67*, 430–450. https://doi.org/10.1111/bmsp.12027

Magis, D. (2015a). *Empirical comparison of scoring rules at early stages of CAT*. Paper presented at the Conference of the International Association for Computerized Adaptive Testing, Cambridge, UK.

Magis, D. (2015b). A note on the equivalence between observed and expected information functions with polytomous IRT models. *Journal of Educational and Behavioral Statistics, 40*, 96–105. https://doi.org/10.3102/1076998614558122

Magis, D. (2015c). A note on weighted likelihood and Jeffreys modal estimation of proficiency levels in polytomous item response models. *Psychometrika, 80*, 200–204. https://doi.org/10.1007/S11336-013-9378-5

Magis, D. (2016). Efficient standard error formulas of ability estimators with dichotomous item response models. *Psychometrika, 81*, 184–200. https://doi.org/10.1007/s11336-015-9443-3

Magis, D., & Barrada, J. R. (2017). Computerized adaptive testing with R: Recent updates of the package catR. *Journal of Statistical Software, Code Snippets, 76*(1), 1–19. https://doi.org/10.18637/jss.v076.c01

Magis, D., Béland, S., Tuerlinckx, F., & De Boeck, P. (2010). A general framework and an R package for the detection of dichotomous differential item functioning. *Behavior Research Methods, 42*, 847–862. https://doi.org/10.3758/BRM.42.3.847

Magis, D., & Raîche, G. (2012). Random generation of response patterns under computerized adaptive testing with the R package catR. *Journal of Statistical Software, 48*(8), 1–31. https://doi.org/10.18637/jss.v048.i08

Magis, D., & Verhelst, N. (2017). On the finiteness of the weighted likelihood estimator of ability. *Psychometrika*. https://doi.org/10.1007/s11336-016-9518-9

Maris, E. (1995). Psychometric latent response models. *Psychometrika, 60*, 523–547. https://doi.org/10.1007/BF02294327

Masters, G. N. (1982). A Rasch model for partial credit scoring. *Psychometrika, 47*, 149–174. https://doi.org/10.1007/BF02296272

McBride, J. R., & Martin, J. T. (1983). Reliability and validity of adaptive ability tests in a military setting. In D. J. Weiss (Ed.), *New horizons in testing: Latent trait test theory and computerized adaptive testing* (pp. 224–236). New York: Academic Press.

McClarty, K. L., Sperling, R. A., & Dodd, B. G. (2006). *A variant of the progressive-restricted item exposure control procedure in computerized adaptive testing systems based on the 3PL and partial credit models.* Paper presented at the annual meeting of the American Educational Research Association, San Francisco.

McKinley, R. L., & Reckase, M. D. (1982). *The use of the general Rasch model with multidimensional response data* (Research Report No. ONR 82-1). Iowa City, IA: American College testing.

Meyer, P. (2014). *Applied measurement with Jmetrik.* New York: Routledge.

Meyer, P. (2015). *Jmetrik (version 4.0.3) [Computer software manual].* Charlottesville, VA: Psychomeasurement Systems, LLC.

Mills, C. N., Potenza, M. T., Fremer, J. J., & Ward, W. C. (2002). *Computer-based testing: Building the foundation for future assessments.* Mahwah, NJ: Lawrence Erlbaum.

Millsap, R. E., & Everson, H. T. (1993). Methodology review: Statistical approaches for assessing measurement bias. *Applied Psychological Measurement, 17*, 297–334. https://doi.org/10.1177/014662169301700401

Mislevy, R. J. (1984). Estimating latent distributions. *Psychometrika, 49*, 359–381. https://doi.org/10.1007/BF02306026

Mislevy, R. J. (1986). Bayesian modal estimation in item response models. *Psychometrika, 51*, 177–195. https://doi.org/10.1007/BF02293979

Mislevy, R. J., & Bock, R. D. (1982). Biweight estimates of latent ability. *Educational and Psychological Measurement, 42*, 725–737. https://doi.org/10.1177/001316448204200302

Mislevy, R. J., & Chang, H.-H. (2000). Does adaptive testing violate local independence? *Psychometrika, 65*, 149–156. https://doi.org/10.1007/BF02294370

Mosteller, F., & Tukey, J. (1977). *Exploratory data analysis and regression.* Reading, MA: Addison-Wesley.

Muraki, E. (1990). Fitting a polytomous item response model to Likert-type data. *Applied Psychological Measurement, 14*, 59–71. https://doi.org/10.1177/014662169001400106

Muraki, E. (1992). A generalized partial credit model: Application of an EM algorithm. *Applied Psychological Measurement, 16*, 19–176. https://doi.org/10.1177/014662169201600206

Muraki, E., & Bock, R. D. (2003). *PARSCALE 4.0 [Computer software manual].* Lincolnwood, IL: Scientific Software International.

Muraki, E., & Carlson, J. E. (1993). *Full-information factor analysis for polytomous item responses.* Paper presented at the annual meeting of the American Educational Research Association, Atlanta.

Nydick, S. W. (2014). *catirt: An r package for simulating IRT-based computerized adaptive tests [Computer software manual].* https://CRAN.R-project.org/package=catIrt (R package version 0.5-0).

Osterlind, S. J., & Everson, H. T. (2009). *Differential item functioning* (2nd ed.). Thousand Oaks, CA: Sage.

Ostini, R., & Nering, M. L. (2006). *Polytomous item response theory models.* Thousand Oaks, CA: Sage.

Parshall, C. G., Davey, T., & Nering, M. L. (1998). *Test development exposure control for adaptive testing.* Paper presented at the annual meeting of the National Council on Measurement in Education, San Diego, CA.

Penfield, R. D., & Camilli, G. (2007). Differential item functioning and item bias. In C. R. Rao & S. Sinharay (Eds.), *Handbook of statistics, Vol. 26. Psychometrics* (pp. 125–167). Amsterdam: Elsevier.

Rao, C. R., & Sinharay, S. (2007). *Handbook of statistics, Vol. 26. Psychometrics*. Amsterdam: Elsevier.

Rasch, G. (1960). *Probabilistic models for some intelligence and attainment tests*. Copenhagen: Danish Institute for Educational Research.

Reckase, M. D. (1979). Unifactor latent trait models applied to multifactor tests: Results and implications. *Journal of Educational Statistics, 4*, 207–230. https://doi.org/10.2307/1164671

Reckase, M. D. (2009). *Multidimensional item response theory*. New York: Springer.

Revuelta, J., & Ponsoda, V. (1998). A comparison of item exposure control methods in computerized adaptive testing. *Journal of Educational Measurement, 35*, 311–327. https://doi.org/10.1111/j.1745-3984.1998.tb00541.x

Riley, B. B., Dennis, M. L., & Conrad, K. J. (2010). A comparison of content-balancing procedures for estimating multiple clinical domains in computerized adaptive testing: Relative precision, validity, and detection of persons with misfitting responses. *Applied Psychological Measurement, 34*, 410–423. https://doi.org/10.1177/0146621609349802

Roskam, E. E. (1987). Toward a psychometric theory of intelligence. In E. E. Roskam & R. Suck (Eds.), *Progress in mathematical psychology* (pp. 151–171). Amsterdam: North-Holland.

Roskam, E. E. (1997). Models for speed and time-limit tests. In W. J. van der Linden & R. K. Hambleton (Eds.), *Handbook of modern item response theory* (pp. 187–208). New York: Springer.

Rudner, L. M. (1998). Item banking. *Practical Assessment, Research, and Evaluation, 6*, 41.

Rulison, K., & Loken, E. (2009). I've fallen and I can't get up: Can high-ability students recover from early mistakes in CAT? *Applied Psychological Measurement, 33*, 83–101. https://doi.org/10.1177/0146621608324023

Samejima, F. (1969). *Estimation of latent ability using a response pattern of graded scores*. *Psychometrika monograph supplement*, Vol. 34 (Monograph no. 17). Richmond: Byrd Press.

Samejima, F. (1974). Normal ogive model on the continuous response level in the multidimensional space. *Psychometrika, 39*, 111–121. https://doi.org/10.1007/BF02291580

Samejima, F. (1977). A use of the information function in tailored testing. *Applied Psychological Measurement, 1*, 233–247. https://doi.org/10.1177/014662167700100209

Samejima, F. (1994). Some critical observations of the test information function as a measure of local accuracy in ability estimation. *Psychometrika, 59*, 307–329. https://doi.org/10.1007/BF02296127

Samejima, F. (1998). *Expansion of Warm's weighted likelihood estimator of ability for the three-parameter logistic model to general discrete responses*. Paper presented at the annual meeting of the National Council on Measurement in Education, San Diego, CA.

Santos, V. D. O. (2017). A computer-adaptive test of productive and contextualized academic vocabulary breadth in English (CAT-PAV): Development and validation. *Graduate Theses and Dissertations, 16292*. Ames, IA: Iowa State University. http://lib.dr.iastate.edu/etd/16292

Schuster, C., & Yuan, K.-H. (2011). Robust estimation of latent ability in item response models. *Journal of Educational and Behavioral Statistics, 36*, 720–735. https://doi.org/10.3102/1076998610396890

Segall, D. O. (2004). A sharing item response theory model for computerized adaptive testing. *Journal of Educational and Behavioral Statistics, 29*, 439–460. https://doi.org/10.3102/10769986029004439

Segall, D. O. (2010). Principles of multidimensional adaptive testing. In W. V. der Linden & C. A. W. Glas (Eds.), *Elements of adaptive testing* (pp. 57–75). New York: Springer.

Smith, R., & Lewis, C. (2014). Multistage testing for categorical decisions. In D. Yan, A. A. von Davier, & C. Lewis (Eds.), *Computerized multistage testing: Theory and applications* (pp. 189–203). New York: CRC Press.

Snijders, T. A. B. (2001). Asymptotic null distribution of person fit statistics with estimated person parameter. *Psychometrika, 66*, 331–342. https://doi.org/10.1007/BF02294437

Stocking, M. L., & Lewis, C. (1998). Controlling item exposure conditional on ability in computerized adaptive testing. *Journal of Educational and Behavioral Statistics, 23*, 57–75. https://doi.org/10.3102/10769986023001057

Stocking, M. L., & Lord, F. M. (1983). Developing a common metric in item response theory. *Applied Psychological Measurement, 7*, 201–210. https://doi.org/10.1177/014662168300700208

Stout, W. (1987). A nonparametric approach for assessing latent trait unidimensionality. *Psychometrika, 52*, 589–617. https://doi.org/10.1007/BF02294821

Stout, W. (2005). *DIMTEST (Version 2.0) [Computer software manual]*. Champaign, IL: The William Stout Institute for Measurement.

Swaminathan, H., Hambleton, R. K., & Rogers, H. J. (2007). Assessing the fit of item response theory models. In C. R. Rao & S. Sinharray (Eds.), *Handbook of statistics, Vol. 26. psychometrics* (pp. 683–718). Amsterdam: Elsevier.

Swaminathan, H., & Rogers, H. J. (1990). Detecting differential item functioning using logistic regression procedures. *Journal of Educational Measurement, 27*, 361–370. https://doi.org/10.1111/j.1745-3984.1990.tb00754.x

Swanson, L., & Stocking, M. (1993). A model and heuristic for solving very large item selection problem. *Applied Psychological Measurement, 17*, 151–166. https://doi.org/10.1177/014662169301700205

Sympson, J. B. (1978). A model with testing for multidimensional items. In D. J. Weiss (Ed.), *Proceedings of the 1977 computerized adaptive testing conference*. Minneapolis, MN: University of Minnesota.

Tate, R. (2003). A comparison of selected empirical methods for assessing the structure of responses to test items. *Applied Psychological Measurement, 27*, 159–203. https://doi.org/10.1177/0146621603027003001

Thissen, D., & Steinberg, L. (1986). A taxonomy of item response models. *Psychometrika, 51*, 567–577. https://doi.org/10.1007/BF02295596

Thissen, D., Steinberg, L., & Wainer, H. (1988). Use of item response theory in the study of group difference in trace lines. In H. Wainer & H. Braun (Eds.), *Test validity* (pp. 147–170). Hillsdale, NJ: Erlbaum.

Thompson, N. A. (2009). Item selection in computerized classification testing. *Educational and Psychological Measurement, 69*, 778–793. https://doi.org/10.1177/0013164408324460

Urry, V. W. (1970). *A Monte Carlo investigation of logistic test models*. Unpublished doctoral dissertation, Purdue University, West Lafayette, IN.

van der Linden, W. J. (1998a). Bayesian item selection criteria for adaptive testing. *Psychometrika, 63*, 201–216. https://doi.org/10.1007/BF02294775

van der Linden, W. J. (1998b). Optimal test assembly of psychological and educational tests. *Applied Psychological Measurement, 22*, 195–211. https://doi.org/10.1177/01466216980223001

van der Linden, W. J. (2000). Constrained adaptive testing with shadow tests. In W. J. van der Linden & C. A. W. Glas (Eds.), *Computerized adaptive testing: Theory and practice* (pp. 27–52). Norwell, MA: Kluwer.

van der Linden, W. J. (2005). A comparison of item-selection methods for adaptive tests with content constraints. *Journal of Educational Measurement, 42*, 283–302. https://doi.org/10.1111/j.1745-3984.2005.00015.x

van der Linden, W. J., Ariel, A., & Veldkamp, B. P. (2006). Assembling a computerized adaptive testing item pool as a set of linear tests. *Journal of Educational and Behavioral Statistics, 31*, 81–99. https://doi.org/10.3102/10769986031001081

van der Linden, W. J., & Diao, Q. (2014). Using a universal shadow test assembler with multistage testing. In D. Yan, A. A. von Davier, & C. Lewis (Eds.), *Computerized multistage testing: Theory and applications* (pp. 101–118). New York: CRC Press.

van der Linden, W. J., & Glas, C. A. W. (2010). *Elements of adaptive testing*. New York: Springer.

van der Linden, W. J., & Guo, F. (2008). Bayesian procedures for identifying aberrant response-time patterns in adaptive testing. *Psychometrika, 73*, 365–384. https://doi.org/10.1007/s11336-007-9046-8

van der Linden, W. J., & Hambleton, R. K. (1997). *Handbook of modern item response theory*. New York: Springer.

van der Linden, W. J., & Hambleton, R. K. (2017). *Handbook of modern item response theory.* New York: Springer.

van der Linden, W. J., Klein Entink, R., & Fox, J.-P. (2010). IRT parameter estimation with response times as collateral information. *Applied Psychological Measurement, 34*, 327–347. https://doi.org/10.1177/0146621609349800

van der Linden, W. J., & Pashley, P. J. (2000). Item selection and ability estimation in adaptive testing. In W. J. van der Linden & C. A. W. Glas (Eds.), *Computerized adaptive testing: Theory and practice* (pp. 1–25). Norwell, MA: Kluwer.

van der Linden, W. J., & Pashley, P. J. (2010). Item selection and ability estimation in adaptive testing. In W. V. der Linden & C. A. W. Glas (Eds.), *Elements of adaptive testing* (pp. 3–30). New York: Springer.

van der Linden, W. J., & Veldkamp, B. P. (2004). Constraining item exposure in computerized adaptive testing with shadow tests. *Journal of Educational and Behavioral Statistics, 29*, 273–291. https://doi.org/10.3102/10769986029003273

van der Linden, W. J., Veldkamp, B. P., & Reese, L. M. (2000). An integer programming approach to item bank design. *Applied Psychological Measurement, 24*, 139–150. https://doi.org/10.1177/01466210022031570

Veerkamp, W. J. J., & Berger, M. P. F. (1997). Some new item selection criteria for adaptive testing. *Journal of Educational and Behavioral Statistics, 22*, 203–226. https://doi.org/10.3102/10769986022002203

Veldkamp, B. P. (2014). Item pool design and maintenance for multistage testing. In D. Yan, A. A. von Davier, & C. Lewis (Eds.), *Computerized multistage testing: Theory and applications* (pp. 39–54). New York: CRC Press.

Veldkamp, B. P., & van der Linden, W. J. (2010). Designing item pools for adaptive testing. In W. V. der Linden & C. A. W. Glas (Eds.), *Elements of adaptive testing* (pp. 231–245). New York: Springer.

Verhelst, N. D., Verstralen, H. H. F. M., & Jansen, M. G. (1997). A logistic model for time limit tests. In W. J. van der Linden & R. K. Hambleton (Eds.), *Handbook of modern item response theory* (pp. 169–185). New York: Springer.

von Davier, M., & Haberman, S. J. (2014). Hierarchical diagnostic classification models morphing into unidimensional 'diagnostic' classification models-a commentary. *Psychometrika, 79*, 340–346. https://doi.org/10.1007/s11336-013-9363-z

Wainer, H. (2000). *Computerized adaptive testing: A primer* (2nd ed.). New York: Routledge/Taylor and Francis.

Wainer, H., Bradlow, E. T., & Du, Z. (2000). Testlet response theory: An analog for the 3-PL useful in testlet-based adaptive testing. In W. J. van der Linden & C. A. W. Glas (Eds.), *Computerized adaptive testing: Theory and practice* (pp. 246–270). Boston, MA: Kluwer-Nijhoff.

Wainer, H., Bradlow, E. T., & Wang, X. (2007). *Testlet response theory and its applications.* Cambridge: Cambridge University Press.

Wainer, H., Kaplan, B., & Lewis, C. (1992). A comparison of the performance of simulated hierarchical and linear tests. *Journal of Educational Measurement, 29*, 243–251. https://doi.org/10.1111/j.1745-3984.1992.tb00376.x

Wang, C. (2013). Mutual information item selection method in cognitive diagnostic computerized adaptive testing with short test length. *Educational and Psychological Measurement, 73*, 1017–1035. https://doi.org/10.1177/0013164413498256

Wang, C., Chang, H.-H., & Huebner, A. (2011). Restrictive stochastic item selection methods in cognitive diagnostic computerized adaptive testing. *Journal of Educational Measurement, 48*, 255–273. https://doi.org/10.1111/j.1745-3984.2011.00145.x

Wang, T., & Hanson, B. A. (2005). Development and calibration of an item response model that incorporates response time. *Applied Psychological Measurement, 29*, 323–339. https://doi.org/10.1177/0146621605275984

Warm, T. (1989). Weighted likelihood estimation of ability in item response models. *Psychometrika, 54*, 427–450. https://doi.org/10.1007/BF02294627

Weiss, D. J. (1983). *New horizons in testing: Latent trait theory and computerized adaptive testing.* New York: Academic Press.

Weissman, A. (2014). IRT-based multistage testing. In D. Yan, A. A. von Davier, & C. Lewis (Eds.), *Computerized multistage testing: Theory and applications* (pp. 153–168). New York: CRC Press.

Wright, B. O., & Masters, G. N. (1982). *Rating scale analysis.* Chicago, IL: MESA Press.

Wright, B. O., & Stone, M. H. (1979). *Best test design.* Chicago, IL: MESA Press.

Xu, X., Chang, H.-H., & Douglas, J. (2003). *A simulation study to compare CAT strategies for cognitive diagnosis.* Paper presented at the annual meeting of the American Educational Research Association, Chicago.

Yan, D., Lewis, C., & Stocking, M. L. (2004). Adaptive testing with regression trees in the presence of multidimensionality. *Journal of Educational and Behavioral Statistics, 29,* 293–316. https://doi.org/10.3102/10769986029003293

Yan, D., Lewis, C., & von Davier, A. A. (2014a). Multistage test design and scoring with small samples. In D. Yan, A. A. von Davier, & C. Lewis (Eds.), *Computerized multistage testing: Theory and applications* (pp. 303–324). New York: CRC Press.

Yan, D., Lewis, C., & von Davier, A. A. (2014b). A tree-based approach for multistage testing. In D. Yan, A. A. von Davier, & C. Lewis (Eds.), *Computerized multistage testing: Theory and applications* (pp. 169–188). New York: CRC Press.

Yan, D., von Davier, A. A., & Lewis, C. (2014). *Computerized multistage testing: Theory and applications.* New York: CRC Press.

Yao, L., & Schwarz, R. (2006). A multidimensional partial credit model with associated item and test statistics: An application to mixed-format tests. *Applied Psychological Measurement, 30,* 469–492. https://doi.org/10.1177/0146621605284537

Yen, W. M. (1981). Using simulation results to choose a latent trait model. *Applied Psychological Measurement, 5,* 245–262. https://doi.org/10.1177/014662168100500212

Yen, W. M. (1984). Effects of local item dependence on the fit and equating performance of the three-parameter logistic model. *Applied Psychological Measurement, 8,* 125–145. https://doi.org/10.1177/014662168400800201

Yen, W. M. (1993). Scaling performance assessments: Strategies for managing local item dependence. *Journal of Educational Measurement, 30,* 187–213. https://doi.org/10.1111/j.1745-3984.1993.tb00423.x

Yen, W. M., & Fitzpatrick, A. R. (2006). Item response theory. In R. L. Brennan (Ed.), *Educational measurement* (4th ed.) (pp. 111–153). Westport, CT: Praeger.

Zheng, Y., Wang, C., Culbertson, M., & Chang, H.-H. (2014). Overview of test assembly methods in multistage testing. In D. Yan, A. A. von Davier, & C. Lewis (Eds.), *Computerized multistage testing: Theory and applications* (pp. 87–99). New York: CRC Press.

Zwick, W. R., & Velicer, W. F. (1986). Comparison of five rules for determining the number of components to retain. *Psychological Bulletin, 99,* 432–442. https://doi.org/10.1037/0033-2909.99.3.432

Index

© Springer International Publishing AG 2017
D. Magis et al., *Computerized Adaptive and Multistage Testing with R*, Use R!,
https://doi.org/10.1007/978-3-319-69218-0

Printed in the United States
By Bookmasters

Printed in the United States
By Bookmasters